DATE DUE

DE 18 99			
MY 2 00			
DE 20 00			

DEMCO 38-296

African Exodus

African Exodus

The

Origins

of

Modern

Humanity

CHRISTOPHER STRINGER
AND ROBIN MCKIE

A JOHN MACRAE BOOK

Henry Holt and Company • New York

Henry Holt and Company, Inc.
Publishers since 1866
115 West 18th Street
New York, New York 10011

Henry Holt ® is a registered trademark of
Henry Holt and Company, Inc.

Library of Congress Cataloging-in-Publication Data
Stringer, Chris, 1947–
African exodus : the origins of modern humanity / Christopher
Stringer and Robin McKie.
p.   cm.
Originally published: London: Cape, 1996.
Includes bibliographical references and index.
ISBN 0-8050-2759-9 (alk. paper)
1. Man—Origin.   2. Human evolution.   I. McKie, Robin.
II. Title.
GN281.S87   1997   96-37718
599.93'8—dc21   CIP

Henry Holt books are available for special promotions and
premiums. For details contact: Director, Special Markets.

First American Edition—1997

Designed by Michelle McMillian

Printed in the United States of America
All first editions are printed on acid-free paper. ∞

5   7   9   8   6   4

*To our families
and our futures*

Ex Africa semper aliquid novi.
(There is always something new out of Africa.)

Pliny the Elder

CONTENTS

Illustrations

Acknowledgments

We would like to thank the friends and colleagues who have contributed to the discoveries and ideas discussed in this book, although they will not all agree with our conclusions. Yoel Rak, Jean-Jacques Hublin, Maryellen Ruvolo, and Walter Bodmer read chapters and provided helpful comments. Robert Kruszynski, Rosie Stringer, and Sarah Mitchell gave invaluable general help, and Irene Baxter patiently typed much of the manuscript. We would also like to thank all those who provided material for the illustrations, particularly Akio Morishima, Ian Tattersall, Barbara West, and the Photographic Unit of the Natural History Museum. Last but not least, we would like to thank the Natural History Museum and the *Observer* for their support.

Preface

For the past few years, a small group of scientists has been accumulating evidence that has revolutionized our awareness of ourselves, and our animal origins. They have shown that we belong to a young species, which rose like a phoenix from a crisis which threatened its very survival, and then conquered the world in a few millennia. The story is an intriguing and mysterious one, and it challenges many basic assumptions we have about ourselves: that "races" deeply divide our populations; that we owe our success to our big brains; and that our ascent was an inevitable one. Far from it: people on different continents are closer evolutionary kin than gorillas in the same forest; Neanderthals became extinct even though they had bigger brains than *Homo sapiens*; while chance as much as "good design" has favored our evolution. We can see evidence of this startling 100,000-year-old genesis not only in the bones of the dead, but in the genes of people alive today, and even in the words we speak. It is a remarkable, and highly controversial narrative that has generated headlines round the world and has been the subject of a sustained program of vilification by scientists who have spent their lives committed to the opposing view that we have an ancient, million-year-old ancestry. The debate, which reverberates in museums, universities, and learned institutions across the world, is one of the most bitter in the history of science. How these events came about and how we learned about our true nature, and our African Exodus 100,000 years ago, is explained by a scientist at the very center of the arguments and a journalist who has closely followed every twist and turn of this dramatic scientific story.

African Exodus

1

The Kibish Enigma

A Personal Introduction by Chris Stringer

All great truths begin as blasphemies.
George Bernard Shaw

I have never been able to trace the source of my passion for fossils. Neither, to their eternal bafflement, could my family. Indeed, it was the basis for some unease that I spent so much of my childhood drawing and painting skulls—scarcely a healthy hobby for a growing boy, after all. I was also famed (if that is the right word) in our family for asking, as a youngster on one of my frequent visits to London's Natural History Museum, if the attendants could give me any old bits of skeleton they didn't need. Fortunately, their rebuff was of such gentleness that my ardor for ancient bones was left undented.

Indeed, such was my eagerness to spend my life among fossils that when I discovered it was possible to study human evolution as a special subject in its own right, I promptly discarded my dearly won place at medical school in order to devote myself to anthropology. The decision horrified my headmaster and my biology teacher, the latter concluding a sorrowful lecture with a prediction that I would never get a job "doing that subject." My parents, to their eternal credit, suppressed their unease—and backed my decision.

I have since had many reasons to be grateful for their support. My mother and father helped me to pursue a career which has involved me in one of the great scientific dramas of modern times: the unfolding of a radical new understanding of our birth as a species and about the source of what are commonly called racial differences. This new theory has, in turn, triggered one of the fiercest, most bitter debates about our origins—a considerable achievement for a field already infamous for polemical divisions and open rivalry. As a result, I have been permitted to be both observer and participant in one of the great intellectual clashes of this century.

At the time of my introduction to the subject, however, I was merely concerned with trying to make a career in it. I graduated from University College London, still imbued with a love of the study of human evolution, and began to plan my Ph.D. research in 1969—in an intellectual climate dominated by a widespread antipathy towards social science, an aftermath of the student troubles of 1968. In addition, none of the "hard science" councils—which distributed cash for medical, scientific, or environmental projects—would take responsibility for research on fossil humans. The subject seemed to fall between all three. Despite the efforts of Michael Day (then at the Middlesex Hospital Medical School), Nigel Barnicot of University College London, and Don Brothwell of the Natural History Museum, I failed to raise the interests of any of these organizations.

By the summer of 1970, there were still no signs of a grant and my prospects of becoming a paleoanthropologist looked bleak. My biology teacher's forebodings about my future occupation were beginning to look ominously prescient. Indeed, I was on the verge of giving up my temporary job at the Natural History Museum to go to a teachers' training college when, at the last minute, a spare Medical Research Council post came up at Bristol University's anatomy department: for a student to conduct research on human evolution. Jonathan Musgrave, a lecturer at the department, phoned Brothwell, who gave him my name. Within a month I was on my way to Bristol.

Jonathan Musgrave had studied Neanderthals, that mysterious, sturdy lineage of human precursors who had lived, died, and been buried in the caves of Europe tens of thousands of years ago, and whose relation to our own species, *Homo sapiens*, underpins our understanding of ourselves and our evolution. In Musgrave's case, he had analyzed their hand bones, using a technique called multivariate analysis that exploits mathematical methods for examining many measurements at once.[1] Two or more different objects—a pair of skulls, for example—are carefully surveyed, and a host of different measurements which mirror their shape are collected. Then, using the complex statistical methodology of multivariate analysis, it is possible to produce an overall measure of how much the two objects differ from each other. It is a powerful technique and I intended to use it on Neanderthal heads to determine just how similar they were to those of the Cro-Magnons, a race of early European members of *Homo sapiens* who thrived about 25,000 years ago and who were very similar to men and women today. As we shall see, this relationship is the source of considerable dispute among scientists today, for at its root lies the resolution of our own origins and nature. Did Neanderthals evolve into Cro-Magnons (and modern Europeans), or did the two represent distinct lineages or even species? I intended to bring an objective—not an emotional—approach to resolve these questions. I would use precise instruments such as calipers and protractors to determine skull height, breadth, and width; angle of forehead; projection of browridge; and dozens of other features to place Neanderthals and Cro-Magnons in their evolutionary context.

This work would shape my life and my career, it transpired—though I did not know this at the time. All that concerned me in the winter of 1970 was getting my hands on fossil skulls—many of which could be found in European museums where they had been gathering dust since the turn of the century. So Jonathan and I began to devise an itinerary that would allow me to visit the maximum number of the most important centers. In retrospect, planning and executing such a trip was only possible with the boldness—and naiveté—of youth. (I was twenty-two years old at the time.) Musgrave had traveled around Europe by train to collect his data, but then he had been investigating hand bones, of which there are relatively few in the fossil record. Skulls are bigger, hardier, and tend not to slip through the paleontologist's net so easily—which means there are far more bits of our predecessors' heads scattered round museums than fragments of their fingers and thumbs. And that in turn meant that I needed to visit many more such learned institutions than Musgrave had. Unfortunately, I raised less than a thousand dollars in travel grants, and was forced to use my ancient Morris Minor car, and camp or stay in youth hostels.

So in July 1971, I left for Europe—having only set foot there twice before: on a brief school trip to Paris and to an Italian seaside resort with my parents. I spoke French passably, but otherwise had to rely on a series of phrase books. Letters were sent out to the museums, but a number, especially in France, East Germany, and Czechoslovakia, failed to send replies. I started in Belgium where I spent my first night in what seemed to be a hostel for vagrants, run by nuns! I studied the Spy skeletons (unearthed in 1886) which were among the first Neanderthal bones to be discovered, and which confirmed that these were the remains of a distinct kind of early human being (and not those of a single anatomical freak as had been suggested). In Germany, I was joined by my girlfriend, Rosie (now my wife), and she accompanied me for much of the rest of the expedition. I examined

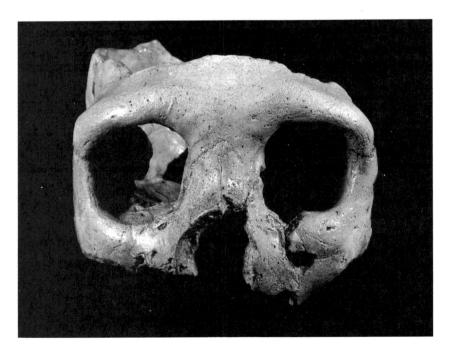

1   The face that launched a frantic glue-hunt! The Krapina fossil which came apart in 1971.

the original Neander Valley skeleton (after which these ancient people were named) in Bonn, and then omitting East Germany on the grounds that a visit there would surely end in tears, I moved on to Czechoslovakia, tense as this was the third anniversary of the Soviet-led invasion. There my car was emptied out all over the road, and I was interviewed for four hours by border officials who made it clear that they found a long-haired, Western anthropology student about as welcome as a hippie at a regimental reunion. "Your visit is of no value to the people of Czechoslovakia," I was told in response to my pleadings that my work was of international scientific importance. I was a tourist, not a researcher, I was informed, and would therefore have to spend ten dollars a day in local currency, a sum that was utterly incompatible with my feeble budget. I cut my losses, and visit, and left Czechoslovakia after five days, although I still managed to find time to study both the Neanderthal and early Cro-Magnon remains kept at Brno.

After a brief stay in Vienna, we moved on to Zagreb, where I discovered that the head of its museum, Dr. Crnolatac, was out of the city and I was refused access to the largest collection of Neanderthals of all—the Krapina fossils excavated by the distinguished prehistorian Dragutin Gorjanović-Kramberger. (His office had been perfectly preserved from that time, I discovered.) Many mysteries still surrounded this material, which had only been studied fully by one Western scientist, Loring Brace, of the University of Michigan, in the previous forty years. Perhaps I looked heartbroken, or possibly suicidal—for a junior curator took pity, and at great risk allowed me access to the Krapina fossils, which were regarded almost as holy relics by the Yugoslav scientific establishment.

I spent three days secluded with these fossil prizes. I was as happy as a fledgling paleontologist could be—until, on the second day, to my horror, one of the most important Krapina fossils, part of a face, came apart while I was measuring it. The prospect of a lifetime of Yugoslavian prison food flashed before my eyes until I realized the skull had separated along an old glued break. Rosie and I raced to a local hardware store, bought a tube of domestic adhesive, and amid much sweaty tension, I carefully reassembled one of the jewels of

Yugoslavian science, watched by the junior curator, saving his neck, my career, and an international incident.

We drove south, with relief, to study another mystery fossil, excavated from the Petralona Cave in northern Greece and stored at Thessaloníka University, before Rosie had to return home, leaving me to take a terrifying drive west across the Pindus Mountains to the ferry terminal at Igoumenitsa. In Italy, I studied the Neanderthal remains from Monte Circeo and Saccopastore, and had my car broken into in Rome, losing a number of important possessions, including a human skull, a recent *Homo sapiens* specimen that I had brought along to act as reference material when making comparisons with ancient skulls—though God knows what my Italian thieves made of it. However, my precious measuring instruments and hard won data were not stolen. If I had lost these, I would—as I recorded in my diary—have simply thrown myself in the Tiber. From then on, I always slept with my data sheets under my pillow—only they were truly irreplaceable.

I finished my tour in France, the country that has the richest collections of human fossils in Europe. Unfortunately, my money had virtually run out, and following a second car break-in in Avignon, I was left with only the few dirty clothes in my laundry bag! I camped in the Bois de Boulogne in late October, with a decidedly limited wardrobe—not the most enjoyable experience of my life. To cap it all, one of the curators at the Musée de l'Homme proved to be less than helpful and I was denied access to the La Ferrassie skeleton, found in 1909 and the most complete of all Neanderthal fossils. Apparently studies of it had still not been completed sixty years after it had been discovered! I was also told that another key fossil skull, from the Jebel Irhoud cave in Morocco, had been returned there, an economy with the truth that was only revealed when secret assistance by another anthropologist—Yves Coppens, whose work on the East Side Story of human evolution will be discussed in Chapter 2—gave me furtive access to it.

At the end of October 1971, I returned home to Bristol after my four-month, 5,000-mile, fossil-measuring marathon. Rosie and I had had our tent swept away during a thunderstorm in Prague, the car

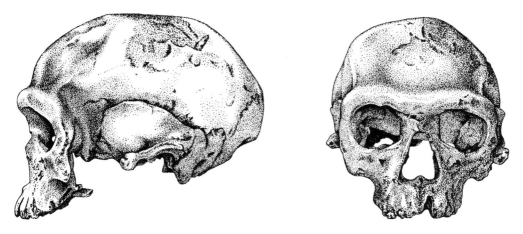

2   A vital African fossil, which Chris Stringer almost missed in 1971—Jebel Irhoud 1, from Morocco.

exhaust had collapsed and had to be held together with a coat hanger, and I had lost fourteen pounds. On the other hand, I was positively fattened intellectually by the firsthand knowledge I had gained of Europe's most important Neanderthal and Cro-Magnon skulls. The next two years were spent analyzing that data, and studying the Natural History Museum's, and other, collections. I devoted much of my time to laboriously transcribing my measurements onto computer punch cards so they could be analyzed by Bristol University's giant computer—a state-of-the-art machine that seemed impressive at the time, but which had less calculating power than a modern desktop machine. By the time I returned to the Natural History Museum at the end of 1973 as a senior research fellow, my Ph.D. was almost finished, and my conclusions were crystallizing.

I was sure—for a start—that Europe did not have two parallel, coexisting lines of human evolution, Neanderthal and modern, as some scientists had argued. Nor was it the case that Neanderthals represented a worldwide stage in the evolution of modern humans, as was also being suggested at the time. It looked like the Neanderthals had evolved in Europe, but that highly dissimilar Cro-Magnons had not. As I completed my Ph.D., I became convinced the Neanderthals were not our ancestors, that the early fossils of Europe recorded their evolution but not that of modern *Homo sapiens*, and

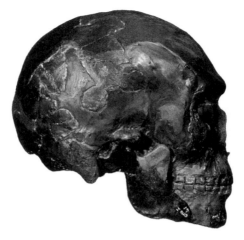

3   Reconstruction of the Omo Kibish *Homo sapiens* skull by Michael Day and
Chris Stringer.

that there was little sign of intermixture between Neanderthals and
early modern people in either Europe or the Middle East, as some
scientists had proposed. The latter had simply replaced the former.
But where did those early modern people come from? I couldn't say
in 1974.[2] There was simply too little evidence.

It was then that I became involved with the man from Kibish.
Strongly built, stained in hues of blue and brown from his lengthy
immersion in the soil, the fragments of his skull, jaw, and skeleton
had been disinterred from their resting place on the banks of the
River Kibish in Ethiopia in 1967. A year later, these anatomical relics
were sent on a brief tour of research centers where they were copied
and measured before being packed and sent back to Addis Ababa
where they have remained ever since. I had first glimpsed these few,
precious fragments of bone in Michael Day's office while still a post-
graduate student, and even included a mention of them in my Ph.D.
However, it was not until 1974, when Michael Day was preparing to
return them to Ethiopia, after measuring and copying them, and
reconstructing their owner, that I got a proper look, an inspection
that was eventually to trigger my fundamental rethink about the evolution of our species.

The man from Kibish (the skeleton's features suggest he was a
male) had a higher and rounder skull and a bigger chin than any

Neanderthal, and his skeleton suggests his was a taller and lighter frame than we find in those archetypal cave people—though he still had a powerful physique compared with an average modern male, with a noticeable browridge over his eyes and a rather broad and receding forehead. His bones had been found by an expedition led by Richard Leakey, who is better known for his discoveries of far more ancient and primitive hominid fossils around Lake Turkana in northern Kenya, only a few hundred kilometers from the Kibish region of Ethiopia. Leakey's own work at Kibish, one of his earliest adventures, was carried out on behalf of his father, Louis, who was leading the Kenyan contingent of a joint French-American-Kenyan dig in the Omo-Kibish area. Its aim was to investigate the sediments lying on either side of the lower reaches of the Omo River where it widens and flows south towards Lake Turkana in north Kenya.

As Richard recalls in his autobiography, *One Life*,[3] he and his team only narrowly avoided being eaten by crocodiles at one stage of their expedition. This was difficult and uncomfortable work; yet, for all their pains, their reward was a meager hoard of human remains: a partial skull and skeleton, a second skull discovered on the opposite bank of the Kibish River, and a third small fragment of skull. No volcanic rocks, which often supply important geological and chronological data, were found around the site. However, samples of shells collected from well above the level of the Kibish dig were dated to around 40,000 years ago, indicating that the bones, which were found far below this sequence of beds, must be far older. In addition, shells from the same level as the Kibish site were dated using a special technique called uranium-series dating—which we shall meet later—to around 130,000 years old. As Leakey put it: "I have probably collected as many fossils as anyone else around, and the one thing I do know is where things come from. The best estimate we could get from the geology and dating . . . would place them [the Kibish fossils] at between 100,000 to 130,000 years old."[4]

At the time, by the standards of other African finds, this seemed an unremarkable date for a supposedly primitive human being. But, beginning with my thesis, I began to think more deeply about that man from Kibish. Surely his skull was too domed, the brow too small, the chin too strong, and the bones too fine for him to be classified as an ancient form of human being? Quite frankly, he basically looked like a modern human and—more than any Neanderthal—he appeared to be a far more likely ancestor for the relatively recent Cro-Magnon people of Europe. Nor was I alone. Michael Day, and a other few researchers whom we shall meet in Chapter 3, were also coming to a similar conclusion. So in 1976, at a Cambridge conference, I presented my ideas on the man from Kibish, and his importance as a prototype modern European. A paper on my lecture was eventually published two years later.[5] However, apart from a few negative reactions such as a fairly dismissive critique of my methods and deductions by Milford Wolpoff, of the University of Michigan, whom we shall encounter later, my theorizing mostly fell on deaf ears.

Michael and I were undeterred, however, and in 1980, in response to an invitation to attend a conference at Nice, we decided to reexamine the Kibish 1 skull, and rigorously test how modern it was. It passed with flying colors. Michael also restudied the rest of the Kibish 1 skeleton and concluded it was equally modern-looking. Indeed, so modern and lithe were its proportions that we realized we might be dealing with a far more revolutionary piece of fossil evidence than we had previously realized. Could this be the oldest modern human of them all, we wondered? Was the Kibish Man a forerunner of all the men and women who populate our planet today? Michael and I presented a cautious assessment of the dating evidence at the conference in 1982 but insisted that the Kibish skeleton did indeed belong to an early modern human, and not a kind of African Neanderthal, as some scientists, such as Brace, had claimed.[6] Our view was supported by Leakey: "Geological investigations and dating have shown that the two skulls [from Kibish] are about 130,000 years old, yet despite their antiquity they are both clearly identifiable as *Homo sapiens*, our own species. At the time of their discovery, scientists generally believed that our species had only emerged in the last 60,000 years and many considered the famous Neanderthal Man to be the immediate precursor to ourselves. The Omo fossils thus provided important evidence that this was not so."[7]

By now I was taking stronger and stronger lines on the distinctiveness of the different populations of early humans, and was edging towards the view that perhaps the Neanderthals were different enough from modern humans to not be grouped with us as *Homo sapiens* at all. (A couple of colleagues based in New York, Todd Olson and Ian Tattersall, and one in London, Peter Andrews, played a pivotal role in convincing me that my data clearly pointed that way. They gave me backbone when I needed it.) Instead, the Neanderthals seemed to represent a long-lived European line of evolution, but one which disappeared from the continent rapidly around 35,000 years ago. Neanderthal characteristics which had evolved and persisted for 200,000 years had been replaced by new ones in less than 10,000 years. I summarized my thinking in 1984 in *Natural History* magazine: "The origin of the Cro-Magnons' modern characteristics seems to lie outside of Europe. Evidence from southwest Asia suggests that modern people replaced the Neanderthals about 40,000 years ago. Before then, it is necessary to look to Africa for the origin of the earliest modern humans."

In other words, the Kibish Man was a far better candidate as the forebear, not just for the Cro-Magnons but for every one of us alive today, not just Europeans but all the other peoples of the world, from the Eskimos of Greenland to the Pygmies of Africa, and from Australian Aborigines to Native Americans. In short, the Kibish Man acted as pathfinder for a new genesis for the human species.

Since then, many paleontologists, anthropologists, and geneticists have come to agree that this ancient resident of the riverbanks of Ethiopia and all his Kibish kin—both far and near—could indeed be our ancestors, though it has also become clear that the evolutionary pathway of these fledgling modern humans was not an easy one. At one stage, according to genetic data, our species became as endangered as the mountain gorilla is today, its population reduced to only about 10,000 adults. Restricted to one region of Africa, but tempered in the flames of near extinction, this population went on to make a remarkable comeback. It then spread across Africa until, by about 100,000 years ago, it had colonized much of the continent's savannas and woodlands. We see the imprint of this spread in biological studies that have revealed that races within Africa are genetically the most disparate on the planet, indicating that modern humans have existed there in larger numbers, for a longer time than anywhere else.

Then there has been the archeological evidence. For example, in the western branch of the African Rift Valley, on the eroded banks of the Semliki River, near the town of Katanda, archeologists led by the husband-and-wife team of John Yellen, of the National Science Foundation, Washington, and Alison Brooks, of George Washington University, have uncovered many treasures, including startlingly elegant bone harpoons and knives. Previously it was thought that the Cro-Magnons were the first humans to develop such delicate carving skills—50,000 years later. Yet this very much older group of *Homo sapiens*, living in the heartland of Africa, apparently displayed the same extraordinary skills. It was as if, said one observer,[8] a prototype Pontiac car had been found in the attic of Leonardo da Vinci.

In addition, the team has found fish bones, including some from two-meter-long catfish. It seems the Katanda people were efficiently and repeatedly catching catfish during the spawning season, indicating that systematic fishing is quite an ancient human skill—and not some relatively recently acquired expertise, as many archeologists had previously thought. On top of this, archeologists have discovered evidence that one of the Katanda sites had at least two separate but similar clusters of stones and debris that look like the residue of two distinct neighboring groupings, signs of the possible impact of the nuclear family on society, a phenomenon that now defines the fabric of our lives.

All this evidence paints an intriguing picture about our recent African forebears and their sophisticated lifestyles. Bands of these people—armed with new proficiencies, like those honed by the men and women who had flourished on the banks of the Semliki—would have begun to thrive and multiply, eventually triggering an exodus from their homeland. Slowly they trickled northward, and into the Levant, the countries bordering the eastern Mediterranean. Then, by about 80,000 years ago, small groups were spreading across the globe via the Middle East, planting the seeds of human modernity in Asia and later on in Europe and Australia. In each region, these seeds slowly germinated until, about 40,000 years ago, something caused them to blossom with an explosive population growth.

It was one of the critical events in mankind's convoluted route to evolutionary success. The nature of the trigger of this great social upheaval is still hotly debated, but remains a mystery at the heart of our "progress" as a species. Was it a biological, mental, or social event that sent our species rushing pell-mell toward world domination? Was it the advent of symbolic language, the appearance of the nuclear family as the basic element of human social structure, or a fundamental change in the workings of the brain? Whatever the nature of the change, it has a lot to answer for. It transformed us, after all, from minor bit players in a zoological soap opera into evolutionary superstars, with all the attendant dangers of vanity, hubris, and indifference to the fate of others that such an analogy carries with it.

Yet the nature of this new trait is not the only baffling question raised by the study of recent human evolution. Today men and women conduct themselves in highly complex ways: some are uncovering the strange, indeterminate nature of matter, with its building blocks of quarks and leptons; some are probing the first few seconds of the origins of the universe 15,000 million years ago; while others are trying to develop artificial brains capable of staggering feats of calculation. Yet the intellectual tools that allow us to investigate the deepest secrets of our world are the ones which were forged during our fight for survival, in a set of circumstances very different from those which prevail today. How on earth could an animal that struggled for survival like any other creature, whose time was absorbed in a constant search for meat, plants, and raw materials, and who had to maintain constant vigilance against predators, develop the mental hardwiring needed by a nuclear physicist or astronomer? This is a vexing issue that takes us to the very heart of our African Exodus, to the journey that brought us from precarious survival on a single continent to global control.

If we can ever hope to understand the special attributes that delineate a modern human being we have to attempt to solve such puzzles. How was the Kibish Man different from his Neanderthal cousins in Europe, and what evolutionary pressures led the Katanda people to make such crucial changes in behavior—in the heart of a continent that has for far too long been stigmatized as backward? These are intriguing questions—though I would never have dreamed my early fossil forays would lead to the posing of such fundamental inquiries.

This book—the work of a scientist, myself; and a journalist, Robin McKie—aims to provide some answers to questions like these, and to help us understand what it means to be human. We shall do this using three main sources of evidence: paleontological (bones), archeological (stones), and DNA (genes)—with a dash of human observation thrown in for good measure. In the next chapter, we shall follow the story of the emergence of the human lineage, a range of creatures who began their separate existence as "apes who left the trees" and who began to walk upright. In the succeeding chapter, we shall explore how studies of our mysterious predecessors, the Neanderthals, have played a powerful role in shaping our ideas about our own evolution, research that led many scientists to conclude that these people were the ancestors of modern Europeans. We shall also focus on the subsequent studies that have led to the overthrow of this notion, an act that has helped trigger a revolution in our self-awareness as a species. In Chapter 4, we will investigate why *Homo sapiens* triumphed at the expense of the Neanderthals. These chapters will reveal how recent are mankind's African origins, a feature highlighted in Chapter 5, which will survey the genetic evidence that underpins this idea and which also shows how startlingly similar is every member of the human population of this planet. Chapter 6 will trace mankind's footsteps as our ancestors embarked on their African Exodus, crossing the world, leaving enigmatic traces of their presence. In Chapter 7, we will investigate the aftermath of that conquest—the establishment of the world's different contemporary populations. These are the various races into which humanity is divided, a classification whose significance has been utterly transformed by our awareness of our recent African origins. Then in the penultimate chapter, we will probe one of the most important questions: Was one piece of mental architecture responsible for our rapid rise to world dominion? And finally, we shall look at the legacy of that ascendancy, the Stone Age bodies and the impulses of modern humans, features that confirm the strange story of our evolution and which ultimately threaten to eclipse us as a species.

Of course, our reconstruction of mankind's history is not the only one that can be created from the scientific data available, but for us it is the most realistic one. As Jacob Bronowski said: "Science is a very human form of knowledge. We are always at the brink of the known, we always feel forward for what is to be hoped. Every judgement in science stands on the edge of error, and is personal. Science is a tribute to what we can know, although we are fallible."[9] Some will therefore disagree with us, and a few may become quite apoplectic, for this is an intensely disputed intellectual area, a scientific battlefield that has witnessed clashes of such vehemence that they make the row

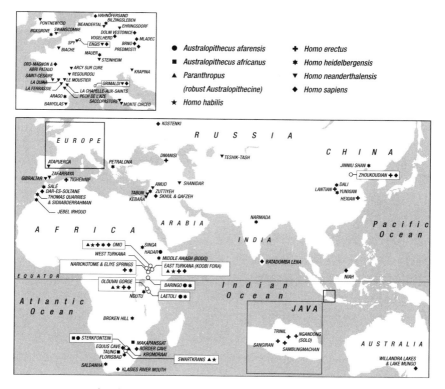

4   Important fossil hominid sites. The African transitional forms to modern humans have been classified here as late examples of *Homo heidelbergensis*.

set off by the Piltdown Man fraud look like an amiable ecclesiastic debate.

These conflicts are extreme because so many cherished notions about our origins have been overturned by the Out of Africa theory. Our book will show why its basic tenets are correct, nevertheless, and will demonstrate that humankind's common and recent ancestry has great importance, for it implies that all human beings must be very closely related to each other (as is also demonstrated by genetic studies). Human differences are mostly superficial, changes which took place in the blinking of an eye in terms of our whole evolutionary history. We may look dissimilar, but we should not be deceived by the stout build of the Eskimo, or the lanky physique of many Africans. What unites us is far more significant than what divides us. Our variable forms mask an essential truth—that under our skins, we are all Africans, the metaphorical sons and daughters of the man from Kibish.

2

East Side Story

Most species do their own evolving, making it up as they go along, which is the way Nature intended. This is all very natural and organic and in tune with the mysterious cycles of the Cosmos, which believes there is nothing like millions of years of evolving to give a species moral fibre and, in some cases, backbone.

Terry Pratchett

Charles Darwin was not infallible. He once conjectured (in early, but not later, editions of *On the Origin of Species*) that bears, which sometimes swim with their mouths open in order to catch insects, might one day evolve into "a creature as monstrous as a whale."[1] He also doubted the rapidity with which multicellular creatures evolved during the Cambrian period 550 million years ago, an evolutionary "explosion" that scientists now take for granted. Such judgments were oddities, however. Darwin's was a mighty, fecund intellect. Given his great output, occasional deviations were inevitable. Far more to the point, when it came to human origins, he invariably hit the bulls-eye—despite a total lack of "hard" fossil evidence to back his hunches. "It is somewhat more probable that our early progenitors lived on the African continent than elsewhere," he wrote in 1881.[2] He was cautious, of course, but nevertheless quite correct.

And if Darwin was alive today, he would have relished all the evidence, derived from sources that range from the bones of the dead to the genes of the living, that has substantiated his inspired guesswork. Equally, he would have been astounded by the variety of fossils of all the other primates, the order of mammals to which we belong, that have also been dug up over the past hundred years. Primates are predominantly long-limbed, tree-loving animals with good vision and dextrous digits, and were so named by Swedish naturalist Carolus Linnaeus to signify our supposed prime position in the animal kingdom.[3] Human beings share this classification with lemurs, marmosets, tamarins, baboons, chimpanzees, gorillas, and many other creatures. Our story is therefore their story, though the rise of *Homo sapiens* (which means, according to Linnaeus, "Wise Man"—a dubious accolade to say the least) is certainly not the simple tale, so often told, of glittering advancement through the primate ranks ending in the inevitable wonder of "the noonday brightness of human genius," as Bertrand Russell described the intellect of our species.[4] As with other creatures, good luck, ill fortune, decline, near extinction, and startling recovery have peppered the story of our evolution. There was nothing predetermined about the human race.

Indeed, its entrance on to the stage of life was distinctly inauspicious. For many millions of years, in an era we call the Miocene, the primate group to which we belong—the apes—had been thriving across the warmer parts of Africa, Europe, and Asia. These large-bodied, tail-less, relatively large-brained animals were a highly successful, widespread, and diverse group. Then they began to die out,

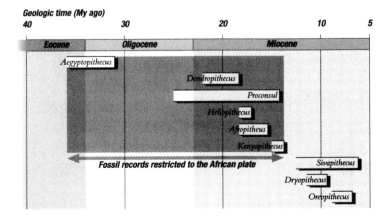

5   Time ranges of the main kinds of fossil apes in Africa and beyond.

losing a battle for resources with monkeys, who tend to be smaller-brained and smaller-bodied, but who nevertheless began to take over the forests of the Old World (Europe, Asia, and Africa) about ten million years ago.[5] The reasons for this shift in the primate power axis are not clear, though anthropologists believe climate change probably played a key role, since the earth began to get cooler and drier then. In addition, some scientists point to the ability of monkeys to digest relatively unripe fruit, a power that would have allowed them to pick off less mature produce ahead of their ape competitors. This is only conjecture, however. What is certain is that apes gradually became isolated in the forests of Africa and southeast Asia, and their numbers have dwindled ever since. Today there are only four other species of great ape (the gorilla, the orangutan, the common chimpanzee, and the bonobo or pygmy chimpanzee), and only the rise of a new group within this primate clan, the hominids (to which *Homo sapiens* and our predecessors, *Homo erectus*, Neanderthals, and others belong) has bucked this decline.

More to the point, some scientists believe the rise and spread of the monkey, and the corresponding entrenchment of the ape, played critical roles in our own evolution. Faced with creatures that displayed greater flexibility in diet and environmental tolerance, some apes began to adapt to life on the level. Our ape ancestors were forced down from the trees, and, once on the ground, evolved upright gait and later the large brains and tool technology that are the distinctive hallmarks of hominid intellect. "According to this interpretation, those few apes that could adapt to a more open, ground-living existence had to develop some decidedly odd features, not in any way 'prefigured' by their initial design—the knuckle-walking of chimps and gorillas, and the upright gait of australopithecines and you know who," says Harvard paleontologist Stephen Jay Gould. It is not a particularly distinguished pedigree, he adds. "Our vaunted ladder of progress is really the record of declining diversity in an unsuccessful lineage that then happened upon a quirky invention called consciousness."[6]

The exact cause, and timing, of the evolutionary split of those apes who elected to remain in or near trees, and who became the ancestors of modern gorillas and chimpanzees, from those who plumped for the plain life, and evolution into hominids, remains a mystery, however. Indeed, researchers who study primate genes are not yet absolutely certain whether chimpanzees are our nearest kin, as seems most likely, or whether gorillas and chimpanzees are equally closely related to human beings. They are sure of one thing though: the immediacy of our relationship with African apes, with whom we share about 98 percent of our genes. Such biological proximity is comparable to that of a zebra and a horse, or a wolf and a jackal—i.e., it is stunningly close, a striking homogeneity that we can see when we study our proteins (whose manufacture is controlled by our genes). Hemoglobin, the oxygen-carrying protein that gives blood its red color, is identical in all its 287 amino-acid subunits with chimp hemoglobin, for example. Simply justifying the separate family name—hominid—which separates humans from the rest of our ape kin is highly questionable scientifically. "Externally, we are so similar to chimpanzees that eighteenth-century anatomists who believed in divine creation could already recognize our affinities," says physiologist Jared Diamond:

> Just imagine taking some normal people, stripping off their clothes, taking away all their other possessions, depriving them of the power of speech, and reducing them to grunting, without changing their anatomy at all. Put them in a cage in the zoo next to the chimp cages, and let the rest of us clothed and talking people visit the zoo. Those speechless caged people would be seen for what they are: a chimp that has little hair and walks upright. A zoologist from outer space would immediately classify us as just a third species of chimpanzee, along with the pygmy chimp of Zaire and the common chimp of the rest of tropical Africa.[7]

In short, the biological abyss that was once supposed to divide human beings from the animals has been revealed to be the narrowest of genetic crevices. Only a 2 percent difference separates the genomes (the collective name for an animal's pool of genes) of human beings and chimpanzees, a wafer-thin discrepancy that is nevertheless responsible for all the wonders of our civilization—from plasma physics and Picasso to pizza. It is an extraordinary phenomenon that again demonstrates how relatively slight, subtle variations in genes and development can still produce profoundly different manifestations in appearance and lifestyle. We shall examine these implications in more detail later in the book.

However, the discovery of this tight genetic bonding has another key implication. If chimp and human are so alike, we cannot have been evolving separately for very long. That momentous split, between those apes who began to walk and those who remained forest dwellers, must therefore be only a very fresh one—about five million years old, as it happens. Yet only a couple of decades ago, those geneticists who first proposed such a recent division invited the derision of most anthropologists.[8] At that time, fragmentary jaw bones and teeth suggested the moment of separation was at least

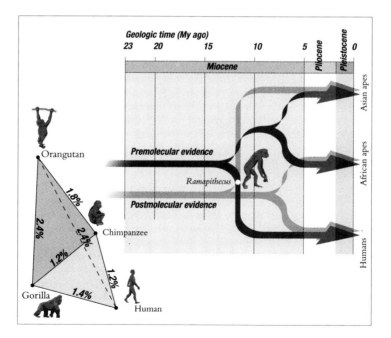

6   How the molecular clock recalibrated our divergence from the apes. Also shown: a network of relationships based on DNA hybridization differences between apes and humans.

three times older. Since then anthropologists have been forced to concede ground (and time) as it has become clear that our supposed ancestor's predecessors, such as *Ramapithecus* from the Indian sub-continent, and *Kenyapithecus* from East Africa, were probably members of more ancient ape lines and that our evolutionary division with the apes is indeed only a very recent one.[9]

The question is: What sort of animal first stepped out of the forest and began a line of evolution that produced creatures that have conquered one planet and who are now exploring the other worlds of the solar system? Until very recently, scientists would only have been able to speculate rather vaguely about its nature. However, this state of ignorance has recently improved in a dramatic way. In 1993, at Aramis, in Ethiopia, an international team of researchers uncovered more than forty fragments—including jaws, teeth, and arm bones— of several primitive individuals with both human and apelike features. These fossil pieces are the remnants of creatures who breathed and walked upon the African landscape about 4.5 million years ago— at a time very close to the ape-human split. The American, Ethiopian, and Japanese anthropologists who found and described the bones gave them a new name: *Ardipithecus ramidus.*[10] This twin nomenclature also harks back to Linnaeus. According to his classification, a living thing—either extant or extinct—is given a first name that recognizes its genus, i.e., the grouping to which all closely related species belong, and a second that records the individual species' name. Hence *Homo sapiens*—Wise Man, or to be absolutely precise, Man the Wise or *Ardipithecus*, the genus name which means "ground ape," and the species name *ramidus* ("root" in the local Ethiopian language) to signify its evolutionary position.

Many scientists believe *ramidus* lies at the root of the human family tree (hence the name), very close to the point where apes and protohumans diverged on their separate paths and destinies (although exactly where and when this divergence occurred is not clear). Certainly, some of its features are clearly humanlike. The canine (eye teeth) were smaller and closer in shape to those of later hominids called australopithecines (literally, Southern apes), and the base of the skull was smaller, a clue to the presence of that most fundamental of hominid characteristics: walking upright, sometimes called bipedalism. Large skull bases support large neck muscles which keep an animal's head from drooping down, as would be needed if it walked on four legs. Dogs and chimpanzees have this feature, for example. A small skull base indicates that the head of *ramidus* probably balanced well on top of the neck and did not need powerful muscles to hold it up, as is also the case with upright human beings. On the other hand, remains dug up with *ramidus*'s bones—including squirrels, colobus monkeys, and other forest-loving creatures—suggest it did not stray that far from its arboreal home. Science therefore still awaits the discovery of other remains to confirm this diagnosis, as one of the team's leaders, Tim White, of the University of California, Berkeley, has acknowledged. "The Aramis remains are anatomically hominid, and they were probably functionally hominid—bipedal—as well. But for now, the evidence from skull, teeth, and arms is all indirect. We are looking for pelvis, knee, ankle, or foot bones that will provide direct evidence," he told *Discover* magazine in late 1994.[11] In addition, bits of skull would indicate the brain size of *Ardipithecus ramidus*—though to judge from later australopithecines, it is unlikely *ramidus* had anything other than a small-ape-sized brain. (White achieved success in his quest in late 1994 when he and his colleagues unearthed another ninety fossils at the *ramidus* site, including a half-complete skeleton. All these precious finds await further study.[12])

On the other hand, some of *ramidus*'s features are apelike: relatively small back (molar) teeth with only a thin enamel covering on their crowns, features found today in gorillas and chimpanzees, who generally eat soft foods (mainly fruits) which require crushing but not a lot of chewing. This was probably the diet of *ramidus*, in this case. In contrast, the large surface area of the back teeth and the thicker layer of resistant enamel of later hominids suggest we changed to food which was either much more abrasive, or needed a lot more chewing, or both, such as nuts, seeds, and tubers.

But if happenstance, in the form of monkey competition, initiated our ancestors' first steps on two feet, a very different set of constraints were to ensure there would be no blacksliding, and no return to a four-legged life in the trees. At that time, the geological forces that had been slowly splitting East Africa from the rest of the continent, creating the majestic Rift Valley, began to make a telling impact on the landscape. Following a line that runs through Ethiopia, Kenya, Tanzania, and finally into Mozambique, two gigantic sheets of rock—known as tectonic plates, upon which continents rest—were slowly moving apart. The fissures and upheavals which this underground division produces have provided a constant background of environmental changes that have played a crucial role in driving evolution in the area. In this case, the seeping magma that leaked out during this great geological parting created blisters of rock and domes of mountains in Ethiopia and Kenya. New lakes and river systems were formed, while mountain ranges—pushed up on either side of the Rift Valley—drastically altered climate continuity across equatorial Africa. While moist forests continued to the west, the east became drier, an effect that was probably intensified by the development of a seasonal monsoon-driven climate in the Indian Ocean, a meteorological effect produced by the rise of the Himalayas. The ancestors of today's African apes contentedly thrived on in the forested center and west, but hominids (whose origins we assume lie in East Africa) became increasingly isolated in the other, less densely forested side of the continent. The mighty rivers and lakes that began running north-south down the Rift Valley only increased the hominids' isolation.

This vision of geological intervention was originally put forward by primatologist Adrian Kortlandt[13] thirty years ago, and has been more recently developed by Yves Coppens,[14] a French paleontologist, who has dubbed the theory "The East Side Story." At its heart, it states that about five million years ago East Africa displayed a flourishing tropical environment, like that in the west. But by four million years ago, after *ramidus* had apparently strutted his (and her) stuff on the ground of East Africa, the ecology began to change thanks to those geological influences. Woodlands and forests became patchier and were separated by more savannas and grasslands. Entirely new kinds of pigs and monkeys entered the scene, and existing species of elephant, rhinoceros, and pig underwent evolutionary change, growing teeth more proficient at chewing, and possessing a greater grazing capacity. These habitat alterations only served to channel hominids further down the road that zoological circumstances had already initiated and which geological forces were now reinforcing. "The population of the common ancestor of humans and apes found itself divided," says Coppens. "The western descendants of these common ancestors pursued their adaptations to life in a humid, arboreal milieu; these are the apes. The eastern descendants of these same common ancestors, in contrast, invented a completely new repertoire in order to adapt to their new life in an open environment: these are the humans." In short, we are "unquestionably the pure product of a certain aridity."

Around this time a new hominid species, *Australopithecus afarensis* appeared on the scene and began its million-year presence in East Africa.[15] (It was called *Australopithecus*—Southern ape—

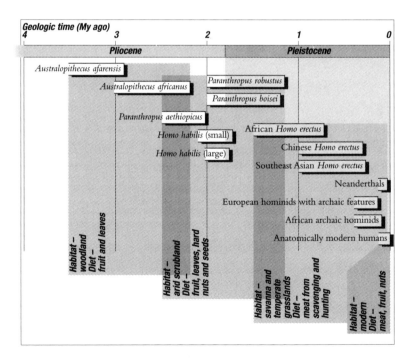

7   Fossil hominids through time, showing probable habitat and dietary changes. The three species of robust australopithecine are assigned here to a distinct genus, *Paranthropus*.

after a South African find we shall discuss shortly and *afarensis* after
the Afar region of Ethiopia where some of the best-known fossils
were discovered.) The species was still an ape like the gorilla and
chimpanzee, in many respects. It was small, about four feet in height,
with short legs and long arms. However, it could clearly walk upright
regularly, a simple act that we now take for granted. Yet its advent
was an evolutionary event of enormous consequence. The first of
three classic features that distinguishes humans from other animals—bipedalism, dependency on toolmaking, and large brains—
had certainly made its appearance by now, if it had not already done
so with *ramidus*. Without it, the other processes which directed our
evolution would probably have taken a very different turn. We should
not, therefore, underestimate *Australopithecus*. As Owen Lovejoy,
an anatomist at Kent State University, puts it: "The move to
bipedalism is one of the most striking shifts in anatomy you can see in
evolutionary biology."[16] It is a view shared by Richard Leakey:

> The origin of bipedal locomotion is so significant an adapta-
> tion that we are justified in calling all species of bipedal ape
> "human." This is not to say that the first bipedal ape species pos-
> sessed a degree of technology, increased intellect, or any of the
> cultural attributes of humanity. It didn't. My point is that the
> adoption of bipedalism was so loaded with evolutionary poten-
> tial—allowing the upper limbs to be free to become manip-
> ulative implements one day—that its importance should be
> recognised in our nomenclature. These humans were not like
> us, but without the bipedal adaptation they couldn't have
> become like us.[17]

It is worth noting that these three critical human attributes did not
arrive simultaneously as a special evolutionary "triple pack" as has
often been suggested in the past. For a long time, it was thought big
brains drove a need to free hands and arms which could then make
tools—which our developing intellects subsequently invented. This
was not the case. Upright stance came first, brains and tools came
later (and not until then can science, in our opinion, justifiably use

the term "human"). But why did we begin to walk upright? We could have stepped out on to the level on four legs just as easily as two. Indeed, that is probably what we did, but then fairly quickly evolved a two-legged gait. So why did we do it this way round? This is one of the most baffling questions in paleontology today, though there is no shortage of suggested answers. Perhaps it was to reach food, or to carry provisions to a home base. Or did early hominids develop an ability to throw stones in order to attack prey, and also for defense? Alternatively, we may have adopted our upright stance to minimize the amount of skin surface exposed to the sun's harsh rays, allowing our ancestors to keep their brains from overheating and to conserve much needed water that would otherwise have been lost in sweat. In other words, our ancestors may have started to walk tall to stay cool, with a two-legged gait reducing the impact of the searing heat that would have otherwise beaten down on their backs. Instead, it would have fallen vertically on their heads—a far smaller area. In addition, upright *afarensis* may have needed to shed the thick pelts that shield other savanna animals from the sun in order to sweat more efficiently, and would therefore have become the first naked ape.

To demonstrate this "cooling hypothesis," Peter Wheeler, of John Moores University in Liverpool, has employed the talents of Boris, a one-foot model of *afarensis* that can be molded into both two-legged and four-legged positions. A camera tracked Boris in his various vertical and horizontal positions, while the movements of the sun over twenty-four hours were simulated. Then scanners measured the size of Boris's photographic image on different frames that represented different times of the day. "We found a very clear pattern," says Wheeler. "We discovered that there was a 60 percent reduction in the heat received by Boris in his two-legged position compared with his quadruped posture—because he presented a much smaller target when the sun was overhead." In addition, in an upright stance, far more of an animal's body is raised above the hot ground and away from the heat radiating from it, while further cooling would have been achieved through contact with breezes and air currents found several feet above ground level.[18]

It is an intriguing theory—though further proof is required, as with all the other suggestions about the origins of bipedalism. Each has its advantages, and its own proponents, as well as its detractors and drawbacks. In short, the jury is still out on the origins of humanity's upright gait.

As for *Australopithecus afarensis*, it displayed some early human-like characteristics. Equally, it possessed features that we would now consider to be strange and atypical of our species. In particular, there appear to have been large differences between the bodies of males (averaging about 4 ft. 10 in., 143 lb. in weight) and females (3 ft. 3 in., 66 lb.), which suggests a very nonhuman, nonmonogamous social structure where males developed large bodies to compete with each other for access to large groups of females—harems; in other words, a social grouping, and a dichotomy of form, that is displayed in other species, such as the gorillas. Females, reciprocally, may have favored larger, "fitter" males. Equally, there is no evidence of toolmaking, and little sign of "intellect," *afarensis*'s brain being ape-sized (between 350–500 ml. in volume, compared with the modern human's average of 1,200 to 1,600 ml.). The *afarensis*'s forehead was only a little more developed than the average ape's, and the face's muzzle a little less prominent. Only the smaller canine teeth and big molars (providing more surface area for better chewing of nuts, berries, and seeds) signaled anything distinctive. The rest of the skeleton, known from finds such as "Lucy," a 40 percent complete skeleton of a three-million-year-old *afarensis*, indicates a mixture of ape and human characteristics: relatively long arms, short legs, and a pyramid-shaped chest like an ape; fairly curved hand bones and short thumbs that could still have formed an effective hook to support the body from tree branches; but comparatively shorter and broader hip-bones that were more like a modern human's.

Afarensis faded from the fossil record about three million years ago, just as a related species made its appearance at the other end of Africa, a hominid that was to trigger some of the most sensational theorizing that has ever been concerned with the issue of human nature. Discovered during quarrying at a lime works at Taung, near Kimberley in 1924, the first remains to be found of this hominid line—the front half of a skull, with jaws and teeth—made their way to Raymond Dart, the newly appointed professor of anatomy at Witwatersrand University, Johannesburg. He gave it the name *Australopithecus africanus* (Southern Ape of Africa), the first time the *Australopithecus* nomenclature was used.[19] Although Dart recognized that adult features had still to develop on the Taung fossil (which, he said, came from a child aged about six years, to judge from the fact that the first permanent molars had just erupted), he was not restrained in claiming the species was a predecessor of modern humans, was intelligent, and made tools. Dart's affirmations were ignored by the British scientific establishment. African australopithecines were only apelike remnants, left behind as human evolution unfolded in Europe and Asia, it was said.

We now know that Dart was right in some ways. *Africanus* has been shown to be much more ancient than the human fossils of Europe and Asia and could perhaps have been an ancestor of ours. However, some of his other assertions went far beyond his meager evidence. He concluded the species was made up of confirmed killers, "carnivorous creatures that seized living quarries by violence, battered them to death, tore apart their broken bodies, dismembered them limb from limb, slaking their ravenous thirst with the hot blood of victims, and greedily devouring living writhing flesh," he wrote in an essay on "The predatory transition from ape to man."[20] This astonishing, almost pornographic outpouring was based on Dart's interpretation of the damaged skulls and bones of *africanus*, and other animal species, found at Taung, and later at Makapansgat and Sterkfontein. He argued that the injuries to both animal and hominid were those inflicted by crude weapons of bone and stone, produced as these ancient killers indulged in orgies of slaughter, and sometimes cannibalism.

Dart's hypothesis was seized upon by Robert Ardrey, an American playwright, who transformed his ideas into a series of sensational bestsellers, starting with *African Genesis*[21] (which inspired, ironically, the title of this book) and which promoted the same notion: that mankind's origins were bloody and violent. Far from evolving big brains and then tools, "the weapon fathered the man," Ardrey claimed. It was our development of stone axes and spears that triggered our

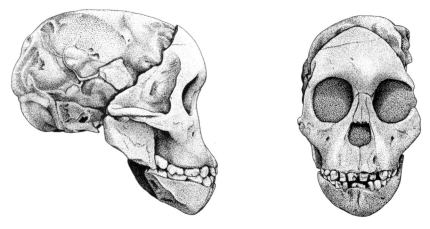

8   The Taung skull—Dart's *Australopithecus africanus*.

evolution to our present status, a process that has been driven by the engines of war. This notion was in turn picked up by Stanley Kubrick and Arthur C. Clarke in their film, *2001: A Space Odyssey*. An ape-man, influenced by unseen aliens, is seen playing with bones. Suddenly he realizes their potential, and he begins to pound the ground, and later his adversaries, with them, before finally flinging a bone weapon into the air where it is transformed into a spinning spacecraft. (In the Monty Python version, the spaceship then spirals back to Earth and squashes the ape-man.) The cinematic image is clear: technology has been driven by our urge to make weapons, and to murder. It was a convenient idea, for it suggests that war is in our genes, and is beyond the control of even the most reasonable men and women. We should feel no guilt or responsibility for our bloody actions, the theory implies. Killing is instinctive and natural.

Yet this whole extraordinary edifice was based on flimsy, and as we now know, misinterpreted evidence. Poor, maligned *africanus* probably did not use tools at all, never mind weapons, and far from being a hunter, was in fact the hunted.[22] Those jumbles of skulls and bones found at Makapansgat, Sterkfontein, and elsewhere had been left by leopards and other predators who had brought their prey, including *africanus* males, females, and children, to their secluded lairs to eat them without being disturbed. The Taung child, for example, is now thought to have been the victim of an eagle which carried off its severed head to a nest.[23] As for the damage marks on the other skulls and bones, these were caused not by human weapons but by predators' teeth, or by the sustained pounding they received as other bits of cadaver or rocks dropped on them. There is no justification for believing that humanity is innately depraved. We extrapolate about our nature from limited data at our peril.

In fact, *africanus* was in many ways like its (possible) ancestor, *afarensis*, displaying a similar body and brain size, although the differences between males and females seem less marked. The canine teeth were further reduced in size; the back teeth relatively larger; the face a little flatter, with more prominent cheekbones; and the hipbones still not forming a pelvic bowl like that of modern humans. As for lifestyle, *africanus* probably spent its time in groups, like most apes today, in which there was no long-term pair bonding. Individuals walked upright and foraged for a mixed but mainly vegetarian diet, though there may have been some meat-eating. However, this would have largely been opportunistic, and of the type displayed today by chimps, which will occasionally kill and eat monkeys that stray too close.

Around this time, about 2.5 million years ago, the climate of South and East Africa became still drier, with patchier rainfall, a change that was most likely related to the growth of the ice caps at the North and South Poles which would have locked up vast quantities of the moisture of the world's weather systems. Many mammals—elephants, horses, and others—evolved higher crowned teeth, indicating a switch to grass, rather than leaf, diets. The hominid line began to change as well, producing three species of "robust" australopithecines, which possessed enormously thick jaws and big back teeth, a specialization that has led many experts to classify them as a different genus, *Paranthropus* ("near man"), as opposed to the "graciles" (*afarensis* and *africanus*). The robust australopithecine brain and body were only a little larger than those of their gracile predecessors, despite a larger face, jaws, and teeth. (Gracile is a less common word for "graceful," and it implies that an individual is lightly built, as opposed to being robust.) They were also bipeds, but still without the capability for long-distance striding and running that is found in *Homo sapiens*.

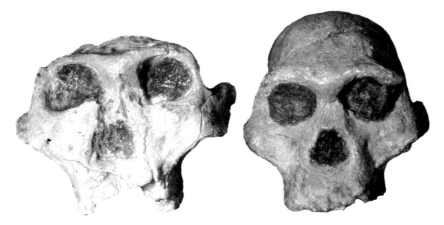

9   Skulls (from left) of a South African "robust" and "gracile" australopithecine, from Swartkrans and Sterkfontein, respectively.

However, robust australopithecines were only one hominid response to the increasing aridity that was continuing to fuel the engine of evolution in East Africa 2.5 million years ago. Their mighty grinding jaws were able to process large amounts of tough plant foods typical of parched environments, and their evolution represented a specialized line of hominids that branched off from the trunk that led to *Homo sapiens*, and which survived up to a million years ago. The second biological reaction to this aridity is far more important to our story and involved the appearance of a creature that took a far more elastic approach to food acquisition. Instead of concentrating its anatomical resources on evolving great grinding mechanisms for crushing hardy vegetation, this species adopted a flexible, broader strategy when it came to gathering food, including meat. This was the first real human, a member of the genus *Homo*, and it appeared in Africa about 2.3 million years ago. It was named *Homo habilis* ("Handy Man") in 1964 by a team led by Louis Leakey (father of Richard), who based his interpretation on finds made in Olduvai Gorge, in the Tanzanian Rift Valley.[24] *Habilis* had smaller jaws and teeth than australopithecines and its skeleton had, supposedly, more humanlike proportions. However, the crucial features were its regular toolmaking and its brain size, about 600 to 750 ml., a clear cut above ape or australopithecine levels. Our ancestors had crossed

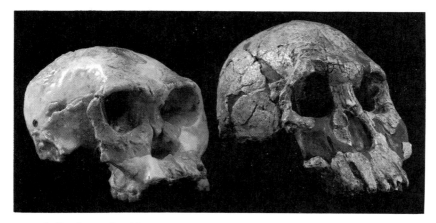

10   Skulls of a small (Olduvai hominid 24) and large (KNM-ER 1470) *Homo habilis.*

their cerebral Rubicon, and had begun the cultural event known as the Lower Paleolithic (sometimes known as the Early Old Stone Age), an intellectual plateau that was to stretch out over the next two million years.[25]

We therefore have a deceptively neat picture, so far. A small-brained apelike creature, *Ardipithecus ramidus*, emerges from the forests five million years ago and begins the evolution to walking upright. *Ramidus* evolves into *afarensis* and then into *africanus*, before the lineage divides, about 2.5 million years ago, into robust australopithecines, and *Homo habilis*, the first hominid to demonstrate a modestly large brain. (We should note, though, that it took at least 2.5 million years from the point when we started to walk upright to when we began to develop significantly bigger brains.) But even if we ignore the numerous uncertainties of this first part of our story, we cannot avoid the developing complexities of the later parts. The two post-*africanus* arms of the tree of humanity have become more than a little confused of late, with discoveries suggesting that by two million years ago several different hominid types were coexisting: robust australopithecines, possibly two separate species of *habilis* (one large, and large-brained, the other small, and small-brained),[26] and the most advanced hominid to have evolved by that date, *Homo erectus.* Fossil evidence indicates each had slightly different lifestyles and diets, and could find a comfortable niche in the mosaic of forests, woodlands, grasslands, and lakeshores that were to be found in East Africa two million years ago. After a few hundred thousand years, first the *habilis* species, and then the robusts died out, leaving *erectus* in sole charge of human destiny—and with a continuing mystery hanging over its origins.

And *erectus* certainly makes a convincing human. Compared with *habilis*, these men and women became sophisticated toolmakers (which is far more, please note, than being a mere tool user: sea otters carry stone anvils to open shellfish, and beavers make dams, for example) and their artifacts are often uncovered where the bones of antelope, pig, zebra, hippo, buffalo, and elephant are also found. These sites frequently occur beside ancient lakes and rivers where animals would have gathered for water and for shelter provided by nearby trees and bushes. Both full-time carnivores—lions and leopards, for example—and humans would have been drawn here. It is therefore impossible to say whether humans or animal predators killed these creatures. Possibly, *erectus* men and women found prey after its death, or scared away its killers. Certainly, compared with the skills of modern human hunter-gatherers, our ancestry seems to be more of a forager and a scavenger, rather than an outright, full-time killer. This interpretation might also explain why *erectus* was big bodied. If it was competing with wolves and vultures for access to carcasses that had been killed and partly picked over by large carnivores, such as lions, a fairly substantial frame would have helped scare off its scavenging competitors.

Whatever the case, the use of tools to obtain meat opened up an entire new ecological niche for humans. For the first time, technology allowed humans to manipulate the environment. "Each new tool opened up possibilities that were formerly the prerogative of very specialized animals," states Jonathan Kingdon in his book *Self-Made Man and His Undoing*:

Where diggers had needed heavy nails, now there were stone picks, cats no longer had the monopoly of sharp claws, spears mimicked horns, porcupine quills or canine teeth and so on. Here for the first time, was an animal that was learning a multiplicity of roles via the invention of technology. An increasing number of animals now had a new competitor that would encroach on at least a part of their former niche. In some cases—perhaps some of the scavengers—the overlap may have been so great that the hominids took over.[27]

Hominids would never be the same again. In this case, a concentrated and nutritious food—meat—would have provided powerful rewards for those best equipped to gain access to it: individuals with the brains to recall a likely source, or to improve a group's performance at hunting or scavenging. On top of that, meat released important metabolic resources. Our powerful digestive systems, that were then needed to process vegetation low in nutrition, were freed of some of the rigors of their work, providing mothers with high quality food for the brains of their developing babies, and continuing neurological sustenance as those infants grew up. "It was not just meat, but fat and bone-marrow that were being consumed, easy-to-digest foods that permitted the development of smaller stomachs which used up less internal energy," says anthropologist Leslie Aiello, of University College London.[28] "The surplus was used to feed our brains, which began to grow significantly at this time. It was a loop. We started to eat meat, got smarter, and thought of cleverer ways to obtain more meat, although learning to obtain other rich, but easily digestible foods, such as nuts, was probably also involved."

Certainly, the human gut is the only energy-demanding organ that is markedly small in relation to body size compared with other mammals, while the brain size is strikingly large. The latter should weigh about ten ounces for a mammal of our dimensions. In fact, the human brain today weighs almost 3 lbs. Similarly, our gut—including stomach and intestines—is about half its expected size. "And small guts are compatible only with high-quality, easy-to-digest food," adds Aiello. We can see this process of digestive diminution begin with *Homo erectus.* In apes and australopithecines the rib cage is shaped like a pyramid that gets larger as you move down the body, to make way for large stomachs and coils of intestines. *Homo erectus* was the first hominid to have barrel-shaped rib cages which open out to make way for the lungs, and then contract over small gut areas. Similarly, we see clear signs of brain enlargement.

Of course, meat-eating does not make all carnivores clever. It was just that in the case of early mankind, it permitted an already smart creature to get even smarter. Until then our brain size was constrained because, as Dr. Aiello puts it: "You cannot have a big brain and big guts. Providing energy—i.e., food—for both would have kept you so busy you would not have had time for reproductive behaviour."

Such a hypothesis does not explain why humans turned to a wider diet in the first place, but it does show why it was successful. "Our ancestors adopted—by chance—a more flexible approach to sustenance, and developed strategies for eating all sorts of different foods—including meat—that helped their brain size break through the limit imposed by their vegetarian diets," says Aiello. "That was our good luck." The actual drive to adopt wider choices of food was "a niche decision." In the face of increasing aridity and desiccation of lifestyle, humans either had to become specialized devourers of vegetation (the path taken by robust australopithecines), or omnivores. By chance, the latter course freed energy that allowed the brain to grow, making us more efficient omnivores. A loop had been created, one that established and rewarded increased intellect. The improved "smartness" that then ensued established the existence of minds capable of complex social tasks and cohesion, the true hallmark of humanity, as well as its counterpart, inspired individuality. This is a very special mental amalgam to which we shall return in greater detail in the last chapters of this book. The crucial point is that we bore this intellectual necklace round the world like a totem, one that opened up environments that had been previously shut to our ape ancestors.

Intriguingly, the first fossils of this omnivorous progenitor of *Homo sapiens* were not found in Africa, but on the Indonesian island of Java. Eugène Dubois, a Dutch doctor, inspired by the evolutionary writings of the German biologist Ernst Haeckel, took a job in the Dutch East Indies in 1887 to search for fossils of the "missing link," and at Trinil, on the banks of the river Solo, found a peculiarly flattened skullcap with a strong browridge above its (missing) eye sockets, and a thighbone which was certainly fully human. Dubois

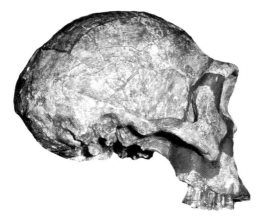

11   A 1.8-million-year-old *Homo erectus* skull from Koobi Fora, Kenya (KNM-
ER 3733).

named the species *"Pithecanthropus erectus"* (erect ape-man). Today
we call it *Homo erectus*, while the Dubois fossil is still often given the
title Java Man.[29] Subsequent discoveries of *erectus* fossils were made
in China (Peking Man) and at various African sites such as Koobi
Fora and Olduvai Gorge. In each case, the skull had thick walls and a
braincase reinforced with bony ridges across the back, top, and sides
(especially in the Asian examples); the eye sockets were dominated
by glowering browridges; and there was a low, or nonexistent, forehead leading on to a relatively long and flat-topped skull. The teeth
were distinctly smaller than those of australopithecines and *habilis*,
but the lower jaw was still thick-boned and chinless.

For the rest of the *erectus* body, we had little to go on, until 1984,
when one of the most spectacular finds in modern paleontology
transformed that state of fossil ignorance in dramatic style. A team
led by Richard Leakey had begun working in a remote area of
northern Kenya in a promising new fossil-hunting area to the west of
Lake Turkana. One member, Kamoya Kimeu, found a small, rather
uninspiring fragment of human skull near the dry bed of the Nariokotome River. Leakey was unimpressed. "Seldom have I seen anything less hopeful," he recorded in his field diary.[30] And so next day
he and his colleagues left Kimeu and traveled to more hopeful paleontological pastures. When they returned that evening they found

that the fragments had magically multiplied and were beginning to show the shape of the skull of a *Homo erectus*. Leakey happily ate his words as, over the next few weeks, the jaws and face, held in the roots of an acacia tree, and most of the other bones were uncovered and the precious skeleton was pieced together.

The bones belonged to a child whose skeleton is the most complete *Homo erectus* ever found. Leaky had uncovered a unique vision of our past, a snapshot of humanity as it existed 1.5 million years ago.[31] Analysis of the pelvis, which differs in shape between males and females even before adolescence, and studies of bone growth revealed the skeleton was that of a boy. The teeth most closely matched that of a modern eleven-year-old (the second molars, for example, had begun to wear down, but the wisdom teeth had only just started to form). However, the size and maturity of the skeleton suggested a more advanced age, nearer fourteen or fifteen. This latter observation indicates a growth pattern somewhat different from that of modern children: one that, like that of apes, lacked the delayed adolescent growth spurt which characterizes *Homo sapiens*. In other words, he was an eleven-year-old with a physique more like a fifteen-year-old modern human's.

To judge from his skeleton, the Nariokotome boy's body shape and size was very similar to that of modern East Africans—tall, long-legged, and narrow-hipped, giving a large skin surface area to assist heat loss in a hot, dry climate by radiation and sweating. Estimates suggest he was about 5 ft. 3 in. tall at death, quite impressive for an eleven-year-old, and implying an equally impressive 6 ft. 1 in. for his adult height. Far from being brutish and short, our predecessor was tall and elegant. He also appears to have been well-fed, as far as we can determine from his sturdy skeleton. The boy was about 78 lb. at death, and would have weighed in at nearly 154 lb. if he had made it to adulthood. His spinal column shows most of the features of a spine of today, but he had an extra lumbar (lower back) vertebra. The shape of the vertebral canal that carries the spinal cord downwards from the brain is distinctive in the way it narrows in the region of the rib cage. This indicates a relative lack of both additional gray matter and enlarged spinal nerves in that region of the spinal column. The

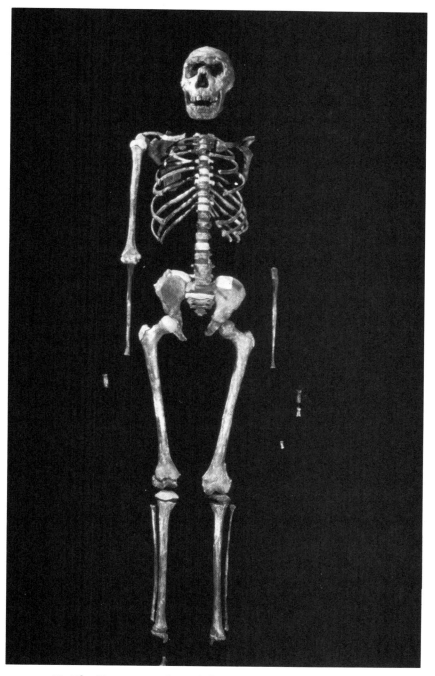

12 The *Homo erectus* boy's skeleton from Nariokotome, Kenya.

Nariokotome boy may have lacked these features because he did not have such good control of his lower trunk or rib cage muscles as modern humans and may therefore not have acquired the very fine breathing control which we employ, quite unconsciously, in everyday speech. Language, as we understand the term, had probably not yet fully evolved. As to his brain, this was about double the volume of a typical ape and about two-thirds of the modern human average. The braincase is still longer and lower than in modern humans, and its shape on the inside displays a difference between the right- and left-hand sides, which in living people is correlated with right-handedness. (In right-handed people, the left back half of the skull is noticeably enlarged because the brain's wiring is crossed-over between its two hemispheres. The left side controls the right side of the body's movements, and vice versa. As a result, people who use their right hands and feet predominantly have slightly larger left rear hemispheres, and vice versa for left-handed individuals. Given that the vast majority of people fall into the former category, this brain shape predominates—a pattern that we see emerging in the days of *Homo erectus*.)

The only clue to the death of the Nariokotome boy can be seen in an inflammation of the lower jaw where one of his milk teeth had recently been shed and which may have resulted in septicemia. Without modern treatment with antibiotics, this is a common cause of death in childhood—so perhaps the lad died of blood poisoning. From the positioning of his skeleton, it seems he fell face down in a marsh where his body was gently spread out as it rotted and broke up. At one stage, his bones were trampled on by large mammals, such as hippos or giraffes, who left their footprints nearby. Later, his remains may have been sucked by catfish and chewed by turtles to judge by the fossils found around him. All were then buried under layers of swamp mud for 1.5 million years until the covering sediments were eroded, exposing some of the fragments—for their fateful meeting with Kamoya Kimeu in 1984.

This then was the form of the first transcontinental traveler, for, as we have seen, *Homo erectus* was the first hominid species that we know was not confined to Africa. Until recently, it was presumed it began an Old World colonization about a million years ago, armed with an ideal physique for long-distance walking, a stone tool technology, and social organization that gave it the power to colonize some fairly harsh and varied terrains—at least from an ape's point of view. However, this fairly simple picture was complicated in 1994 when a team of American scientists dated two Javanese *erectus* skulls at 1.6 to 1.8 million years ago, estimates that approach the age of the earliest Kenyan *erectus* fossils and which suggest the species must have been extremely nippy in its African exodus, or that its predecessor—as yet unknown—did the job first. Either way, if the Java fossils are that old, they reopen the question of where *erectus* originated.

Regardless of the issue of its exact point of origin, *erectus* had clearly begun to spread round the Old World's warmer regions by one million years ago, where local populations developed their own individual features, and may have reached Europe (to judge from a lower jaw, found recently at Dmanisi in neighboring Georgia[32] and fragments of head and body parts found at Atapuerca in Spain)[33] more than 800,000 years ago. However, some experts remain cautious about these early finds, for the continent's cooler climates and longer winters would have presented serious barriers to colonizers lacking the sophisticated social behavior and tools possessed by later human colonists. The problem would have become even more severe about 700,000 years ago as the world's climate machine (which had been running gradually cooler for several million years) became locked into 100,000-year cycles of short, warm, relatively moist weather periods (of the type we are experiencing today), punctuated by much longer stretches of colder, drier conditions— i.e., the Ice Ages. For all that, the descendants of *erectus* did eventually make their definitive mark in Europe during this glacial parry and thrust. A thick and chinless lower jaw found at the Mauer sand quarry near Heidelberg, Germany, in 1907, is believed to be about 500,000 years old. Similarly at Boxgrove, near Chichester (a site already noted for its well-preserved, sophisticated stone implements and butchered animal bones), a massive human shinbone was discovered in 1993.[34] It belonged to an individual nearly as tall

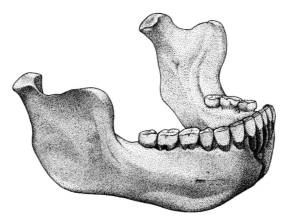

13   The Mauer mandible, type specimen of *Homo heidelbergensis.*

as the Nariokotome boy would have been had he reached adulthood, but who was probably even heavier (at least 168 lb.). The
bone is strong, with thick walls, indicating it had to withstand constant heavy usage.

Once established both in Africa and Europe, these "archaic
sapiens"—as they are known to many scientists to denote their
halfway-house status between *erectus* and modern *sapiens*—continued
to evolve, with an average brain size increasing to about 1,300 ml.
(about the modern average), a rise that occurred in various human
populations round the world. The question is: Why? What pressures
continued to fuel the engine of brain expansion? There are several possible explanations. Social groupings may have become more complex,
requiring more brain power to map more involved, extensive relationships. Talking or gossip may have evolved as a social "glue," though it is
not thought that complex language as we know it had yet developed.
Indeed, cultural change was painfully slow, to judge by the evidence of
stone tools whose style remains static for hundreds of thousands of
years. As Desmond Clark of the University of California at Berkeley
put it: "If these ancient people were talking to each other, they were
saying the same thing over and over again."[35]

The braincase of archaic *sapiens* was also relatively higher and
more filled out above the ear region, until—by about 300,000 years
ago—this hominid line began to manifest the features of a new

species in Europe, as we can judge from the fossil treasure trove of
Atapuerca, a Spanish cave that has produced the largest single collection of ancient human bones from anywhere in the world.[36] About
1,300 bones and teeth representing the jumbled skeletons of a least
thirty-two men, women, and children were uncovered in a small
chamber deep in the cave, at the bottom of a fifty-foot vertical sinkhole. How these bones were deposited in the cave, we do not know.
All that has been determined to date is that they show an intriguing
mixture of ancestral (*erectus*) and more advanced characteristics.
Some had brain sizes above the modern average, some well below,
and the form of the bones on the side of the braincases looks remarkably modern. The many teeth found at the site are quite small by
erectus standards, and some of the front ones (the incisors) show fine
scratches where something (meat? fibers?) was held in the jaws and
cut with stone tools (held by right-handers). The jaws were less
strongly built than those of *erectus*, but were still chinless. The Atapuerca skeletons have still to be assembled but may well turn out to
be shorter on average than their immediate predecessors. More
importantly, their finger, arm, hip, and leg bones resemble those of
the people who came after them in Europe—the Neanderthals,
those enigmatic hominids who (with other species who evolved from
erectus elsewhere in the world) form the penultimate chapter in the
writing of the Book of Humanity. Quite simply, the Atapuerca people

14   Juan-Luis Arsuaga (left) and Stephen Aldhouse-Green in the "Pit of the Bones"
at Atapuerca, the site which has produced over 1,300 human fossils.

look like primitive Neanderthals and provide a link between the *erectus* lineage and the fully developed European hominids who lived during the last Ice Age.

Meanwhile, in Africa, descendants of *erectus* also continued to evolve and by 200,000 years ago gave way to more advanced humans known as "late archaic *sapiens*," represented by fossils from sites such as Jebel Irhoud in Morocco and Florisbad in South Africa. Skull and face shape were still similar overall to that of African and European archaic *sapiens* and the primitive Neanderthals of Europe, like those from Atapuerca. But browridges were getting smaller, foreheads higher, and the shape of the braincase changing subtly toward a more modern pattern. These people probably retained the taller, narrower-hipped anatomies of their African ancestors, though there are no skeletons complete enough to be sure.

Elsewhere in the world, the fossil record for this period can best be described as patchy. Apart from a piece of skull in Israel, and a battered braincase discovered in India, the paleontological cupboard is virtually bare in west and southwest Asia. China is more revealing, however, and suggests that *erectus* may have lingered long after archaic *sapiens* had evolved into new forms in the west of the Old World. This may explain one of the most puzzling features of the archeology of the Paleolithic (the Old Stone Age). Apart from simple flakes used as knives, the hand axe was the most ubiquitous stone tool across half the inhabited Old Stone Age world, appearing at sites as disparate as Boxgrove and Olduvai Gorge. But these tools never became established in the Far East, possibly because bamboo, not stone, formed the main constituent for tools there, and these have simply not survived the passage of time.

However, recent discoveries have complicated this neat scenario. Two crushed skulls, found at Yunxian and tentatively dated at around 350,000 years old, were revealed to have much larger braincases than *erectus*.[37] In fact, they most resemble the archaic *sapiens* finds of the West. The discovery suggests *erectus* was not the only occupant of the region at this time, an idea reinforced by later Chinese fossils— such as the 200,000-year-old partial skeleton from Jinniushan and a skull from Dali—which are definitely not *erectus* remains (the braincase shape is too advanced). Either their owners evolved locally, and rapidly from *erectus*, or are examples of archaic *sapiens* who migrated into the region. The latter idea suggests population dispersals must have been taking place across Asia by this time, an important notion that we shall discuss in greater detail in the next chapter.

Even more puzzling was the discovery, in 1936, from deposits of the Solo River at Ngandong, of twelve braincases and a couple of leg bones that look unmistakably like those of *erectus* people. Incredibly, recent dating suggests they may be less than 100,000 years old.[38] If so, members of the Solo Man population, as the Ngandong folk are also known, must have been the last survivors of a species which began its reign when the australopithecines and *Homo habilis* still thrived on the savannas of East Africa two million years ago, and which had clung to life at the fringes of the inhabited world, while brash new hominids were evolving elsewhere. However, not every scientist interprets the Ngandong evidence in this way, as we shall see.

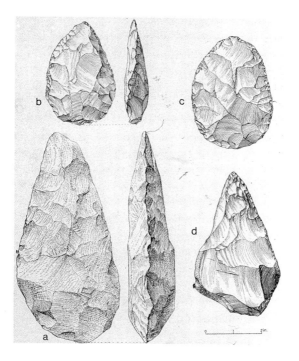

15   Hand axes from Africa (a), the Levant (c), and Europe.

16   The skullcap found in the Neander Valley in 1856.

For the final strands of our brief history of the rise and fall and rise of the hominids, we must return to Europe, however, and especially to the Neanderthals, who have been the most troublesome people of prehistory (for scientists, at any rate) right from the time they first became known to the scientific and popular worlds in 1856, and despite the fact that we know more about them than any other line of ancient humans. Their existence was exposed to the modern world by workmen quarrying in the Feldhofer cave in the Neander Valley near Düsseldorf in Germany in 1856.[39] The men were working on the last few small limestone caves in the valley, and after blasting one open, began to dig out layers of mud, rock, and flint. Then their tools struck bone. First, a skull was unearthed, then thighbones, part of a pelvis, a few ribs, and some arm and shoulder bones. The skull and thigh were particularly distinctive, the former carrying a low, glowering browridge, while the latter was thick and curved. The workmen thought they had found a cave bear's skeleton but fortunately mentioned their discovery to local schoolteacher Johann Karl Fuhlrott, a keen natural historian who recognized their value, though he could scarcely have recognized their true importance—for mankind's self-image would never be the same again.

The uncovering of these remains came at a significant time in the development of ideas about human antiquity. The old concepts of a young Earth and a stable creation were crumbling under the assaults of a wealth of evidence being gathered or analyzed from around the world by a new generation of botanists, zoologists, geologists, and paleontologists. Charles Darwin had returned many years earlier from his voyage around the world in the *Beagle* and was drawing together his notes and thoughts for *On the Origin of Species*, which was finally published in 1859. Charles Lyell, a friend of Darwin, was one of several geologists who had laid the groundwork for an acceptance of a relatively ancient origin for humans, building on the work of a series of prehistorians such as Charles Frère and Boucher de Perthes who had argued for a considerable antiquity for the stone tools (paleoliths) that were being found in ancient river and lake deposits, and in caves.

So the world was ready for Neanderthal Man (the Man of the Neander Valley). Yet right from the beginning, views were highly polarized about its nature. Was this a genuine ancient inhabitant of Europe, who provided a clue to our own evolutionary history? Or was this a deviant, diseased throwback who had nothing to do with our past? The latter view was taken by the famous German pathologist, Rudolf Virchow, who argued that the peculiarities of the Neanderthal skeleton were due to the bone disease rickets. The German anatomist Mayer went even further and argued that the skeleton's bowed leg bones marked him as a horseman and that his damaged elbow showed he had been injured in battle. He had therefore been, most probably, a Cossack cavalryman who had penetrated Prussia in 1814 in pursuit of Napoleon's retreating army, had suffered a sword injury, and had crawled into the cave to die. (The fate of his horse, sword and uniform, and the fact that he had apparently been buried, were all conveniently ignored.) The agony of his final days caused him to screw up his face in pain, leading to the growth of enormous browridges above his eyes! The explanation may be nonsense, but at least it serves to give a taste of things that were to come as scientists battled to make sense of this extinct line of humans.[40]

In the end, it was left to Thomas Huxley to point out the primitive, but nevertheless human, characteristics of the skullcap, and for William King, an Irish anatomist, to propose that this was an ancient human, biologically different from us. On this basis, he named the first distinct ancient human species *Homo neanderthalensis*. Intriguingly, in doing so, King robbed the Rock of Gibraltar of a possible prior claim to a role in the species nomenclature. In 1848, the skull of a Neanderthal woman had been blasted from a quarry, but after

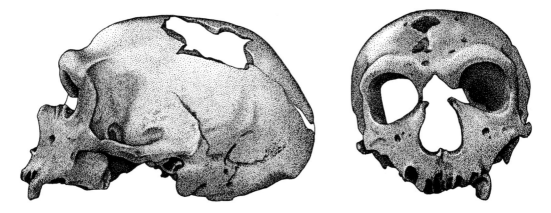

17 The skull of the "Old Man" of La Chapelle-aux-Saints, described in great detail
by Marcellin Boule.

some initial local interest in it, it was neglected, until George Busk displayed it at the British Association for the Advancement of Science meeting in Bath in 1864. Hugh Falconer wrote him a rather lighthearted letter with a serious point, that this skull was distinctive enough to represent a new species of human: *Homo calpicus* (named after an ancient name for the Rock of Gibraltar—Calpe). But as this name was never properly published in the scientific literature, the Neanderthal find got all the scientific priority, attention, and controversy.[41]

As the nineteenth century progressed into the twentieth, more Neanderthal-like remains were found in caves, particularly in Belgium and France, and it soon became clear that these could not all represent Cossack horsemen or diseased individuals. Marcellin Boule, the eminent French paleontologist, described one of these finds, the skeleton from La Chapelle-aux-Saints, in great detail and his publication was to exert perhaps the greatest influence on scientific thought about the Neanderthals for the next half century.[42]

Boule recognized the essential humanity of the La Chapelle man, but was puzzled by its peculiar mixture of primitive and advanced characteristics. Lacking our present knowledge of the earliest stages of human evolution, Boule tried to push the La Chapelle man into the position of a kind of "halfway-house" between apes and humans, giving him toes which could grasp, and a bent-kneed shuffle of a walk. On the other hand, he realized the skeleton could not be very ancient in geological terms, while features like the large brain (indicated by the volume of the braincase) and prominent nose were clearly not primitive. Boule has been heavily criticized in recent years for his mistakes of interpretation, particularly for not taking into account the effects of disease (such as arthritis) on the skeleton. Yet his errors were due as much to his ignorance about the previous course of human evolution (entirely excusable, given the fossil hominid record known at the time) as to his lack of knowledge about the extent of modern human anatomical variation (less excusable). He eventually concluded that the Neanderthals were an offshoot of the main line of human evolution that had retained apelike features but had also developed its own specializations, paralleling, or even exceeding, those of modern people.

At the same time, other research on Neanderthal fossils was being carried out but was largely ignored in the French and English-speaking world. (A very large collection of early Neanderthal fossils had been excavated and described by Dragutin Gorjanović-Kramberger from the Croatian site of Krapina around the turn of the century, for example.)[43] It took many more years before these other finds were properly integrated with the growing body of data from western Europe.

These early discoveries still map out only part of the Neanderthal range in time, space, and anatomy, however. The finds from Neander, La Chapelle, and Gibraltar show us the best-known late western European Neanderthals of the last Ice Age, whereas the Krapina specimens represent an earlier eastern variety. Significant additions to the range of the Neanderthals in this century have come from further east, as far as the cave of Teshik-Tash in Uzbekistan, over two thousand miles from the Neander Valley, and in the Middle East, with the finds from Shanidar in Iraq and Tabun, Kebara, and Amud in Israel. (At Skhul and Qafzeh in Israel, anthropologists and paleontologists found other remains that have been interpreted as showing a mixture of features of both Neanderthals and more modern-looking people. We shall discuss their importance in subsequent chapters.) However, as far as we know, there were never Neanderthals in Africa or the Far

East, since these regions were inhabited by different kinds of people with their own distinctive features and evolutionary histories.

Nor should we think of the Neanderthals as lacking sophistication. They did improve on the stone-crafting techniques of their predecessors, and had developed some more efficient toolmaking techniques and specialized instruments to add to their trusty, million-year-old, vintage hand axes. This cultural stage is known as the Middle Paleolithic or Middle Old Stone Age.

In Europe, Neanderthals were succeeded about 35,000 years ago by people who are named after the cave at Cro-Magnon in France, one of the first sites at which their bones were discovered, in this case in 1868. The Cro-Magnons had higher, more domed skulls, with small browridges and prominent chins. They were taller, and longer-legged, and although they were still fairly muscular, with relatively big teeth, the walls of their leg bones were thinner than those of Neanderthals, and other earlier hominids. In other words, they looked quite a bit like people today.

Cro-Magnon remains are invariably found with tools of the type known as Upper Paleolithic or Late Old Stone Age. These were often made up of long, thin blades of stone, struck off in large and economical quantities from specially chosen cores, and then modified at their ends or sides to create specialized knives, scrapers, piercers, and engravers. To this varied tool kit, Cro-Magnons added the first intensively worked pieces of bone, ivory, and antler—materials which had been strangely neglected until then, despite their ubiquity. These were made into beads, delicate needles, and other objects. However, to many people, the Cro-Magnons' crowning achievement was their art. Not only did they make engravings and sculptures, but they also modeled in clay, and—most spectacularly of all—they covered the walls of deep, subterranean chambers with vivid images of deer, horses, bison, mammoths, and other contemporary animals. To date, such ocher- and soot-daubed murals have been found in more than two hundred caves in western Europe, with 90 percent being found in three regions: the Bay of Biscay on the coast of northern Spain, the foothills of the central Pyrenees, and, the biggest cluster of all, within a twenty-mile radius of the village of Les

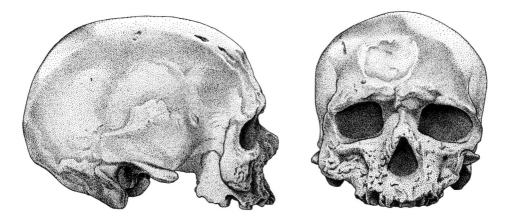

18   The skull of the "Old Man" of Cro-Magnon, discovered in 1868.

Eyzies in the French Dordogne. Some of these are relatively old, such as the 33,000-year-old "vulva" markings (believed to be symbolic representations of female pudenda) that cover the walls of La Ferrassie in France, while others, such as the magnificent animals that are painted on the walls of Altamira cave in Spain, are relatively recent, being only about 12,000 years old. (Indeed, these are almost the last flourishes of this great artistic outpouring, for by about 11,000 years ago, Cro-Magnon cave painting disappears from the archeological record.)

However, of all of these great sites, the Lascaux cave, in France, is easily the most striking. Indeed, prehistory "has left no record more spectacular," says John Pfeiffer, author of *The Creative Explosion*,[44] an exploration of modern humans' artistic origins which begins with a singularly striking description of the Lascaux galleries:

> It is pitch dark inside, and then the lights are turned on. Without prelude, before the eye has a chance to become intellectual, to look at any single feature, you see it whole, painted in red and black and yellow, a burst of animals, a procession dominated by huge creatures with horns. The animals form two lines converging from left and right, seeming to stream into a funnel-mouth, toward and into a dark hole which marks the way into a deeper gallery.

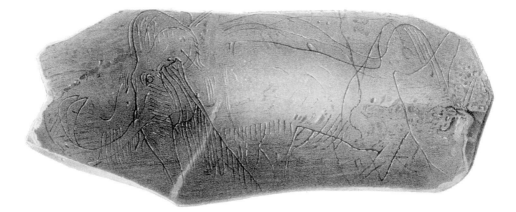

19 A mammoth, carved on a mammoth tusk, from La Madelaine. One of the
earliest discoveries of Cro-Magnon art, found in 1864.

What is so striking about Lascaux and the other caves is that these
truly astonishing works of craftsmanship and imagination seemingly
erupt without precedent. There is little to foreshadow their emergence, no sign of clumsy, crude beginnings, though (as we shall see
when discussing early *Homo sapiens* in Africa and Australia) it is very
unlikely that people did not make some form of picture, perhaps on
hides or other perishable materials, or on their own bodies, before
then. However, it was during the Upper Paleolithic in Europe, and
equivalent periods elsewhere, that humans began creating works of
art that endured, demonstrating symbolic thought and a profound
creative expression. In this elegant handiwork, we can, for the first
time, identify the signature of a creature that is truly like ourselves,
people who were beginning to make their mark on their environments
in a very different and lasting manner, and who were clearly under-
going a cultural revolution of critical importance, one that will be discussed in greater detail later in this book.

For the moment, however, we must be concerned with a different
issue—the origins of these Cro-Magnon artists, and for that matter,
all those other groups of early *Homo sapiens* for whom we find evidence appearing round the globe at this time. Did they evolve, in the
Cro-Magnons' case, from Neanderthals who were their immediate
predecessors in Europe? Or should we look elsewhere, following the
words of Boule? As he writes, in the 1946 edition of his *Les hommes*

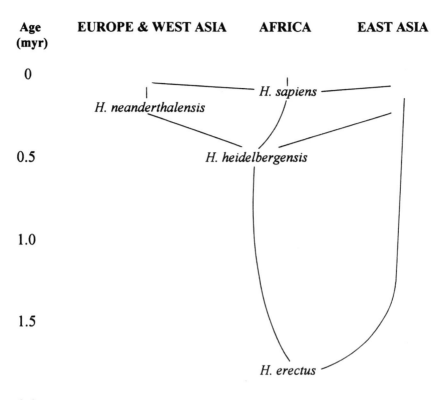

Age (myr) | **EUROPE & WEST ASIA** | **AFRICA** | **EAST ASIA**

20 A simple representation of human evolution over the last 1.5 million years. Some scientists classify the earliest African *Homo erectus* as a more primitive species called *Homo ergaster* ("Work Man"). See B. Wood, chapter 23 ref. 27. Here the archaic *sapiens* forms are regarded as representing two distinct species: *H. heidelbergensis* and *H. neanderthalensis*. Multiregionalists would not recognize the evolutionary lineages and species as distinct—all could be regarded as representing one interlinked species, *Homo sapiens*.

fossiles: "These Cro-Magnons, which seem to replace Neanderthals abruptly in our country, must have lived before then in another place, unless we are willing to propose a mutation so great and so abrupt as to be absurd."[45] These were prescient words, for as we shall see in the next few chapters, these Stone Age Michelangelos did indeed come from "another place"—much to the great chagrin, and even distress, of a good many scientists.

3

The Grisly Folk

I suggest that it was the fate of the Neanderthal to give rise to modern man, and, as frequently happened to members of the older generation in this changing world, to have been perceived in caricature, rejected, and disavowed by their own offspring, *Homo sapiens*.

Loring Brace

The multiregional hypothesis is dead. It is dead because it is unproductive, it is uninteresting, and it is wrong.

Clark Howell

"Nothing of his face was visible but a mouth bordered by raw flesh and a pair of murderous eyes. His squat stature exaggerated the length of his arms and the enormous width of his shoulders. His whole being expressed a brutal strength, tireless and without pity." Not, on the whole, a particularly appetizing companion, you might think. Nor would the following individual seem especially enticing. "Hairy or grisly, with a big face like a mask, great brow ridges and no forehead, clutching an enormous flint, and running like a baboon with his head forward and not, like a man, with his head up, he must have been a fearsome creature."

Obviously, few of us would relish such encounters—yet, according to the authors of this descriptive prose, these were the very visions that met the eyes of our Cro-Magnon antecedents each time they encountered one of their Neanderthal cousins. The first unflattering word-portrait comes from J. L. Rosny-Aines in the novel

La Guerre du Feu,[1] written in 1911 (and subsequently filmed in 1981 under the title *The Quest for Fire*); the second is the work of H. G. Wells, in his 1921 short story, "The Grisly Folk."[2] Both were works of fiction, of course. Nevertheless, the authors' emphasis on the apparently murderous proclivities and atavism of the Neanderthal reflected a general reaction to the species that was typical of early-twentieth-century attempts to find places for them, and for Cro-Magnons, in the evolutionary history of modern humans. And if the latter were the harbingers of "civilization," then the former had to be assigned some far more basic role in the scheme of things, it was thought. They had to be distanced from men and women today, and so were turned into remote and unsavory evolutionary relations. Neanderthals were deemed to be brutal and stupid, a view that became entrenched in popular culture with the publication of Marcellin Boule's analysis of the La Chapelle Neanderthal skeleton, a study which concluded that the species was an evolutionary dead end. Boule's great prestige and the intense anti-German feelings of the years following the First World War combined to suppress discussion of the idea, then supported mainly by German-speaking academics, that Neanderthals might have been direct ancestors of *Homo sapiens*. Victims of such reactions included Gorjanović-Kramberger, whom we met in the previous chapter and who maintained that the Krapina Neanderthals represented a logical ancestral stage for modern humans, and Gustav Schwalbe, who argued the same point from a more general biological perspective. They both wrote in German and paid the price in having their work shunned. Not for the first time, nor the last, did mankind view the past through lenses discolored with visions of the present.

In fact, other European researchers had already begun to swim (albeit quietly) against this tide, though—intriguingly—they tended to make their impact far from their own homelands. For example, Aleš Hrdlička, a Czech, became one of the founders of paleoanthropology in the United States, while Jewish, German-born Franz Weidenreich made his name by heading the institute that organized the "Peking Man" excavations in China, before settling in New York. (He had earlier been forced to flee Nazi Germany, after which he hardly wrote another word in German.) Both considered Neanderthals to be ancestors of modern humans.[3]

Weidenreich, in particular, took a broad view of human evolution. His studies in China led him to the idea that each of the world's inhabited regions had its own local lines of human evolution. "At the very appearance of true hominids there must have already existed several different branches, morphologically well distinguishable from one and another, which all proceeded in the same general direction with Mankind of today as their goal," he wrote in 1943. These different evolutionary lines interbred with each other, although they did not all evolve at the same rate. For example, he argued that Australian bushmen "are less advanced human forms than the white man; that is, they have preserved more of the simian stigmata." Weidenreich believed a line could be traced from Chinese *Homo erectus* to the modern Oriental humans; another from early to late *Homo erectus* in Java and on to present-day native Australians; while an African line of descent was more difficult to determine because there was less fossil evidence to establish continuity. As to the Neanderthals, Weidenreich had no doubts. "What we call Neanderthal man is a widely spread evolutionary phase which may well have perished in one circumscribed territory, but had flourished, expanded, and been transmuted somewhere else, and so have given origin to *Homo sapiens*."[4] In total, Weidenreich painted a clear image of human evolution, a picture that suggested that deep-seated racial characteristics—established when *Homo erectus* settled in the Old World more than a million years ago—divided the peoples of the world. The differing features we see today are the hallmarks of their ancient lineages. In other words, those large noses in Europeans, flat faces in Asians, and flat foreheads in Australians can be traced back to the days of *Homo erectus*.

Weidenreich died in 1948, but his work was taken up by a disciple, Carleton Coon, who dedicated his *The Origin of Races*, published in 1962, to the German anatomist.[5] The book was astonishingly comprehensive—every human fossil known to Coon was described, compared, and assigned a place in his global evolutionary scheme—and was hailed as a truly great academic work, at first. "A milestone in the history of

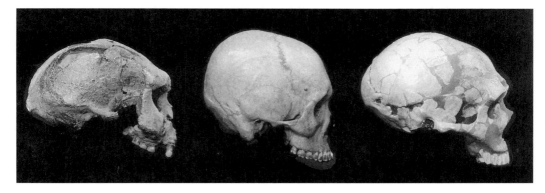

21   Skulls (from left) of a *Homo erectus* from Java, a modern *Homo sapiens* from
       Indonesia, and a Neanderthal from La Ferrassie, France.

anthropology," "a masterpiece," "a landmark," said fellow anthropologists and scientists, who included Julian Huxley and Ernst Mayr.[6]

In *The Origin of Races*, Coon adopted all Weidenreich's arguments and then exaggerated them for good measure. White and Oriental populations were simply more advanced than those from Africa and Australia, he said. As Coon put it: "If Africa was the cradle of mankind, it was only an indifferent kindergarten. Europe and Asia were our principal schools." In his view *Homo erectus* evolved into *Homo sapiens* "not once but five times, as each subspecies, living in its own territory, passed a critical threshold from a more brutal to a more sapient state." But these transitions (which he judged primarily by brain size) did not occur in concert. The *sapiens* level was reached about 250,000 years ago in Europe and Asia, while "the Australian aborigines are still in the act of sloughing off some of the genetic traits which distinguish *Homo erectus* from *Homo sapiens*," he claimed. Coon's attitude to modern human variation was illustrated tellingly in an extraordinary caption that accompanied the last photographs in his book in which an Australian Aboriginal woman and a Chinese man were described as "the Alpha and Omega of *Homo sapiens*." Some academics might have been impressed by Coon's paleontological erudition, but the book's subtext stank of racism to many others.

Coon was attacked with particular savagery by the distinguished geneticist Theodosius Dobzhansky. "There are absolutely no findings in Coon's book that even suggest that some human races are superior or inferior to others in their capacity for culture or civilization," he said. "There are, however, some unfortunate misstatements that are susceptible to such misinterpretation. Professor Coon . . . makes his work susceptible to misuse by racists, white supremacists, and other special pleaders." (Dobzhansky's review was originally commissioned by the *Saturday Review*, who thought it so defamatory, they refused to publish it. Only later did it surface in *Scientific American* and *Current Anthropology*.) And in a subsequent reply to Coon, Dobzhansky wrote that he deplored Coon's refusal to distance himself from the misuse of his book by racists.[7/8] (There is no evidence that Coon, a well-off New Englander with a "genteel, Anglo-Saxon sense of superiority" was an overt racist, though—as Erik Trinkaus and Pat Shipman note in *The Neanderthals*—his habit of referring to people in terms of their racial origin in everyday conservation "led many to believe he attached prejudicial judgements of worth or value to these terms."[9] Others have even stronger recollections, however, including this writer. "In 1979, I (Chris Stringer) went to Harvard for a spell of teaching where I met Coon for the only time just two years before his death. Introducing myself to him in—of all places—the toilets during a break in a seminar, he asked quite calmly: 'And how is that fucking Jew Weiner?,' a reference to the distinguished scientist who had helped expose the Piltdown hoax, but who had reviewed Coon's books in unflattering terms. I was rendered speechless.") On scientific grounds, Dobzhansky also pointed out that it was surely highly unlikely that the transition from *erectus* to *sapiens* would have happened five times independently. The row effectively put an end to Coon's career as a mainstream, respected paleoanthropologist, for he had already taken early retirement from his academic position at the University of Pennsylvania in the year his "great" work was published, and he became increasingly marginalized, even shunned.

The evolutionary schemes of Weidenreich and Coon were consigned to an intellectual limbo, until, in 1977, Alan Thorne, of the Australian National University, presented a paper in which he outlined his Centre and Edge Theory.[10] This resurrected the two scientists' views, and attempted to explain how modern humans and races evolved in different parts of the world over the past million years. Thorne was later joined by Milford Wolpoff, of the University of Michigan, and Wu Xinzhi, from the Institute of Vertebrate Paleontology and Paleoanthropology in Beijing, and they published the definitive version of what they termed "Multiregional Evolution" in 1984.[11] Their account concentrated on the Chinese and Australian fossil records, utilizing many of Weidenreich's observations. "The very earliest Chinese fossils, which are at least 750,000 years old, differ from their Javan counterparts in ways that parallel the differences between northern and southern Asians today," they wrote.

> These folk tend to have smaller faces and teeth, flatter cheeks and rounder foreheads. Their noses are less prominent and are flattened at the top. This combination is also evident in fossils from the Zhoukoudian Cave, the site where the celebrated Peking Man was discovered. Researchers have uncovered specimens there with large brains and other features confirming that the ancient population of China was evolving in a modern direction. Again, various details, such as the shape and orientation of the lower border of the cheek bone, link these fossils with the modern people of the region.

The multiregionalists also owed a clear intellectual debt to Coon, yet although his work was referenced it was hardly acknowledged—because he, much more than Weidenreich, had emphasized the separateness, and apparent sluggishness of some evolutionary lines in their progress toward modern *sapiens* status. Nevertheless, these authors, like Coon, still linked modern human characteristics which they perceived, such as Australians' projecting faces and flat foreheads, with ancient fossil features—in this case the faces and foreheads of their local Javanese *erectus* predecessors. And while Weidenreich argued that such changes arose through an inbuilt drive for evolutionary progress (called orthogenesis), in contrast to Coon, who believed natural selection by and large steered this course to global uniformity, the multiregionalists proposed a different mechanism to account for mankind's status today. They argued that a combination of cultural progress and regular crosscutting gene flow—interbreeding—kept local lineages evolving at the same rate, creating the "glue" that prevented divergence and speciation. As they put it:

> The pattern of modern human origins is like several individuals paddling in separate corners of a pool; although they maintain their individuality over time, they influence one another with the spreading ripples they raise (which are the equivalent of genes flowing between populations).[12]

According to the multiregionalists, our archaic hominid ancestors—*Homo erectus*—emerged from Africa about one million years ago and spread round the Old World. Then, in all the planet's diverse, inhabited corners, islands, remote highland areas and valleys, these early humans slowly evolved in separate and dissimilar ways to produce Eskimos, Pygmies, Australian Aborigines, and all the other manifestly diverse peoples that populate the earth today. For example, in Europe, *Homo erectus* evolved into Neanderthals, which then evolved into modern Europeans.

Such a theory would suggest, at face value, that modern humanity's constituent races are divided by fundamental and deep-rooted differences. However, the multiregionalists argue that genes flowing between populations, "the results of an ancient history of population connections and mate exchanges that has characterised the human race since its inception," offset this trend and have kept mankind on a single path towards its current status. "Human evolution happened everywhere because every area was always part of the whole," add Thorne and Wolpoff.[13] In other words, gene flow ensured that the world's population headed towards the same general evolutionary goal, *Homo sapiens*, and did not deviate down individual, local routes—though it is also claimed that local selective pressures would have produced some distinctive regional physical differences (such as the European's big nose). All the early human types of the ancient world—"Java Man," "Dali Man," "Rhodesian Man," "Solo Man," and "Neanderthal Man"—were therefore part of our collective human ancestry as their genes were constantly shuffled together like cards in a global pack of human evolution, although

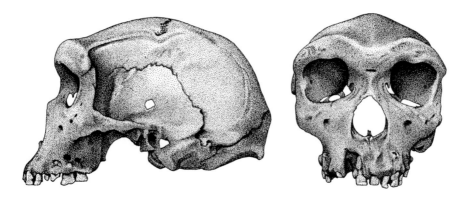

22 The Broken Hill skull ("Rhodesian Man") found in 1921, an African example of *Homo heidelbergensis*.

they also held the seeds of modern "racial" variation because some cards stayed put through all the shuffling. (This vision of mankind's evolution being a history of endlessly shuffling its genes around the globe will be discussed in more detail later.)

This hypothesis has its bizarre consequences, however, ones that have recently forced the multiregionalists to push back the date of origins of *Homo sapiens* to well over one million years—in order to accommodate some of the most disquieting features of their theory. The people usually classified as *Homo erectus* were all, in fact, early *Homo sapiens* according to this new view of human evolution. The origin of *erectus* (presumably from someone like *Homo habilis*) was really the origin of our own species, they say. In their view there were no splits in human evolution from 1.5 million years ago to the present day, so there was only one human species, *Homo sapiens*. While any expert can easily distinguish *erectus* fossils from those of modern *sapiens*, the multiregionalists argue that these differences are only minor variations within an evolving single species. The fact that an example of *Homo erectus* like the Nariokotome boy, unearthed by Richard Leakey and Kamoya Kimeu in 1984 and which we discussed in Chapter 2, had no chin and a brain that had only two-thirds the average volume of ours, is just a minor detail of little significance, it is implied. The "real" fact is that there was only one evolving human population over the last million years or so, and therefore there should only be one name for it—a neat piece of redefinition that helps sidestep the problem of those different evolutionary rates which so derailed Coon. It does not matter if one racial group was slow in developing modern-looking features 250,000 years ago. They already were members of *Homo sapiens*, so the problem merely becomes a minor evolutionary matter of change *within* a species, not one between species. Just like that.

Other scholars are deeply unconvinced by this cavalier reclassification of 1.5 million years of human prehistory, not least Philip Rightmire, author of *The Evolution of Homo erectus*.[14] As he puts it:

> Lumping distinctive populations like Neanderthals and other archaic humans, such as those represented by skulls from Broken Hill, together with *Homo erectus* and suggesting there were no important extinctions across the entire Old World during this period is not going to help us to explore the patterns of evolutionary change that ultimately produced populations like us.

Nevertheless, in their own peculiar way, the multiregionalists do raise an important question: Just how, exactly, is a living species defined? Is it a matter of bureaucracy or consensus or are there solid rules to guide us? In fact, a species is usually designated as a group of organisms which normally do, or could, interbreed to produce fertile offspring (that is, ones which can in turn successfully breed). Closely related, but different species, may interbreed, but either this is not their normal behavior, or the hybrid offspring cannot reproduce in the long term—such as the mule, the sterile offspring of a male donkey and female horse. Even when they meet, closely related species show differences of behavior, physical appearance, or smell, which deter interbreeding. In other words, observation of behavior provides the general (but not hard and fast) rule for classifying living animals, with genetic studies often giving important assistance.

However, if that is true, how on earth do we deal with fossil creatures, and in particular ancient humans? How can we judge whether *Homo erectus* represents a different species from *Homo sapiens*, when we cannot tell if they would naturally and successfully have interbred with us? We have no genes, no flesh, no hair, no sweat to guide us—only bones and teeth. All we can do is look at their skeletons. Unfortunately these give no indication of important differences in, say, mating behavior, that leave no trace in the fossil record—so we are now completely ignorant about those features. There are many distinct monkey species which, if reduced to their bones and teeth, are impossible even for experts to tell apart from one another, for example.[15] So assigning a species name to a set of fossil remains can often be imprecise—and controversial.

But then so is the business of assigning to what race a person belongs. Many people, when asked to give an example of a human race, mention groupings like Jews, Pakistanis, or Chinese. These are cultural or national groupings, or course, and have little biological meaning. And while it is true we would not confuse an average inhabitant of Pakistan with an average inhabitant of China, in most cases we would find it impossible to tell apart a Pakistani from an inhabitant of Bangladesh or India, based only on their physical appearance. The same goes for differentiating between Chinese, Korean, or Japanese. The question is even more difficult to deal with when dealing with Jewish people—for Jews who have lived for generations in Europe look distinctly European, compared with Jews born in Morocco, who look decidedly North African, while the Falasha Jews of Ethiopia look remarkably like other Ethiopians. Indeed, the Falashas look so African that it took a decision by the highest religious authorities in Israel to confirm they really were Jews, when many were airlifted from troubled Ethiopia in the 1980s and early 1990s. (The Falashas claim Jewish origin as descendants of Menelik, the alleged son of King Solomon and the Queen of Sheba, and lived a segregated life in villages in northwest Ethiopia, observing all the traditions of Judaism: the Sabbath, monogamy, circumcision, and biblical laws of purity. The name Falasha is Ethiopian for stranger, and during the civil wars and famines that afflicted Ethiopia in the 1970s and 1980s, they suffered great hardship until thousands were airlifted to Israel in a rescue operation sponsored by the Israeli government.)

In fact, assigning race can be a highly subjective act of categorization. In the United States, one study—carried out in the early

1970s—found that 34 percent of participants in a census in two consecutive years changed racial groups from one year to the next. "Race is supposed to be a strictly biological category, equivalent to an animal subspecies," says Yale anthropologist Jonathan Marks. "The problem is that humans also use it as a cultural category, and it is difficult, if not impossible, to separate those two things from each other."[16]

Nevertheless, the issue of race is crucial to the study of human origins, or to be more accurate, an understanding of our evolutionary roots holds invaluable lessons for our comprehension of the relations of different living peoples with one another. A look at how scientists have studied different ways of recognizing human "races," based on inherited physical characteristics, is therefore important to our examination of the origins of modern humans. Traditionally these have been the most obvious characteristics of skin color, hair type, and physique (though as we shall see in Chapters 5 and 7, the science of genetics has thrown a whole new light on such classifications and their values). Linnaeus, the first great classifier, made a simple four-way division of the species he created—*Homo sapiens* (leaving aside two spurious categories he also named for "Wild Boys" and "Hairy Men"). His varieties were *americanus, europaeus, asiaticus*, and *afer*, a categorization based on the inhabited continents, but with some physical and behavioral features also included in the description. Thus skin color was introduced (red, white, pale yellow, and black, respectively), and some less defensible behavioral attributes (for example Asians were melancholy and ruled by beliefs, Africans were sluggish and ruled by emotions, while Europeans, naturally, were confident and ruled by laws!).[17]

Johann Friedrich Blumenbach, the German naturalist, modified Linnaeus's scheme: Europeans and western Asians became "Caucasian," east Asians became "Mongolian," and Africans became "Ethiopian." He also created a category for the people of southeast Asia, Polynesia, and Australia—Malay. The Blumenbach system was based on a theory of racial origins—the original, most perfect, form of human was the Caucasian, as supposedly found in Georgia, while the other races had deviated away from this primeval and ideal

23 A Kurdish woman, an example of the so-called "Caucasoid" race.

state.[18] For his part, Carleton Coon argued, in 1962, that there were five human subspecies: the Australoid, Mongoloid, Caucasoid, Congoid, and Capoid (the last two corresponding to the main black African populations, and the "Bushman" or "Khoisan" people of southern Africa respectively.)[19]

By definition, Caucasoids are supposed to have little skin pigmentation (melanin), so their skin, eye, and hair color may be pale (though this is clearly not the case for some Caucasoid populations around the Mediterranean or in India). Their head hair is usually relatively fine and straight, and there is often well-developed body and facial hair in males. The nose is generally narrow and prominent. Mongoloids usually have a pale or light brown skin color, a narrow flat nose and flat face, with prominent cheekbones, dark straight coarse hair on the head, but little body or facial hair in males compared with Caucasoids. The eyes usually have an extra fold in the upper eyelid (called an epicanthic fold). Native Americans are usually recognized as a subtype of the main Asian Mongoloids, though their skin color is less varied and they tend to have more prominent noses.

Negroids, or Congoids, are supposed to have dark or black skin, eyes, and hair color; lips which look thick because they are turned outwards from the mouth; a broad, flat nose; and woolly hair. The Capoid race defined by Coon is supposed to have flatter faces with more prominent cheekbones, a paler brown skin color, tightly curled

24   Eskimo women from Labrador—"Mongoloids."

hair, and eyes which sometimes have epicanthic folds. The Australoids have well-pigmented skin, though some individuals, particularly as children, may have fair hair. Body and facial hair is usually
well developed in males. Faces and noses are broad, and males may
have narrow foreheads and quite strong browridges.

Such descriptions are very simplified, of course, and the different
groups rarely have sharp borders between them when they overlap in
a habitat. Indeed, many populations simply do not fit into any of
these categories. Take the Ainu, an aboriginal population from Japan.
Their ancestors apparently occupied the islands long before other
Japanese arrived there. Ainu tend to have long brown hair, prominent noses, round eyes, and the males have much more facial and
body hair than other east Asian people. Some anthropologists claim
they are a peculiar form of Mongoloid, others that they are Caucasoid, and some that they are Australoid.

But why do Caucasoids and Mongoloids and the other racial categories, used by Coon and the rest, look so different in the first place?
Why does a man from northern Europe have blond hair and blue
eyes, and a woman from equatorial Africa have dark skin and tightly
curled black hair? And what produces the spectrum of human variation in between? Well, the most obvious cause is natural selection.
Many generations of exposure to extreme environments results in
physical changes. For example, very hot climates favor narrow, cylin-

25   Three West Africans—"Negroids."

drical bodies with large surface areas of skin that can radiate off heat and keep individuals cool and healthy. In higher latitudes, where the weather is cold, surface area has to be minimized to conserve heat— so people evolve more spherical body shapes. (A sphere has the lowest surface area in proportion to its volume, compared with any other object. Rounder animals therefore have less exterior from which heat can radiate—and so stay warmer). The explanation there- fore accounts for the contrast between the physiques of Kenyan tribesmen and the native residents of Greenland and Lapland. Evo- lution works on the variation in every population so that, over many generations, people with physiques best suited to their environment will thrive and produce more children, who in turn would inherit their parents' successful physiques.

Then there is the question of skin color. In areas of strong sunlight, dark skin provides protection against dangerous ultra- violet radiation that can cause skin cancer. But in the gloomier parts of the world, such as northern Europe, dark skin interferes with the formation of vitamin D in the skin, which plays a critical role in metabolizing calcium for the bones in our bodies. A lack of it leads to bone deformation, however, particularly to rickets. Afflicted with this condition, our early forbears would have been left at a dis- tinct disadvantage during the day-to-day business of foraging and hunting. In addition, women with rickets would have suffered pelvic

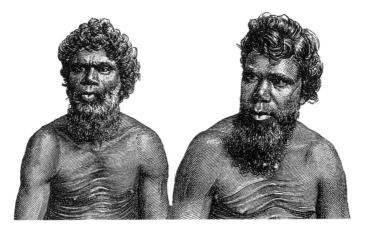

26   Two men from New South Wales—"Australoids."

deformations and would have faced extremely high risks of death during childbirth. This could have produced an intensive selective pressure on early Europeans and other high altitude dwellers, one that would have ensured that the genetically determined attribute of light skin evolved fairly quickly.

(Individuals with light skins who now live in sunny climes pay the price for this adaptation today by suffering greatly increased risks of getting skin cancer. White Americans, who are particularly prone to sun worship, are seven times more likely to get skin cancer than black Americans living in the same area. A similar phenomenon is seen in sun-loving Australia. In northern Queensland, one in ten white men between the ages of sixty and sixty-nine now has skin cancer, and deaths from melanomas are the highest in the world.)[20]

The relationship between skin color and levels of sunlight is not an exact one, however. For example, the now extinct Tasmanian Aborigines, whose climate was like that of northwest Europe, had skin color that was nearly as dark as their tropical, native Australian relatives who are used to far sunnier weather. On the other hand, natives of the Andes do not have skin that is especially dark, even though they have to endure some of the highest ultraviolet radiation levels in the world.

Clearly other factors also play a part in selecting human character- istics—and one of these is sexual selection. Different societies have different views about what is an ideal physical appearance and this can skew populations. Some may consider body and facial hair to be unattractive, while others might favor it. In the latter case, hairier individuals are slightly more likely to find partners and reproduce successfully, thus gradually accumulating genes for body hair for future generations. The same may be true of skin color, with darker or lighter varieties being preferred in different societies.[21]

Isolation and chance can also play a role in shaping the human form. A population that has been cut off from the rest of the world will be deprived of contact with other populations that might moderate differences which may be developing in their physiques or appearance. This process is called genetic drift. In addition, when a new land is settled by an atypical small extract from a population, this can produce dramatic skewing as well. If this little group of founding colonists had, on average, slightly lower foreheads than the population from which they were drawn, this difference could be multiplied across a whole, previously uninhabited land. This phenomenon is known as the founder effect.

And this brings us back to the theory of multiregional evolution which, as we have seen, proposes that this racial variation is the result of long-term regional developments. In each area, populations developed characteristics from the time that *Homo erectus* first evolved (perhaps in Africa two million years ago) or arrived there (in Java and China, over a million years ago, in Europe more than 500,000 years ago). They developed these features as a result of isolation and in response to local environmental conditions. The world's different peoples began as primitive humans, quite different from us in bodies and, probably, behavior, and soon began to develop "racial" features once they had spread across the world. They began to look more like modern humans only over the last 200,000 years or so, seemingly changing at different rates in different parts of the planet.

Now it is important to consider the issues of species and race from the point of view of the multiregionalists. Racial differences are a consequence of populations—regionally isolated from one another in distinct conditions—evolving local quirks and characteristics suiting these local conditions. This isolation and development continued

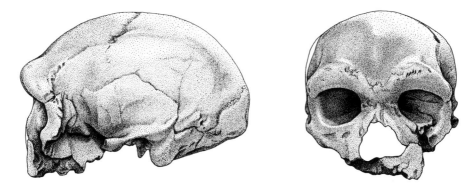

27   The Dali skull. Perhaps a late Chinese example of *Homo heidelbergensis*. For multiregionalists, this could be a direct ancestor of modern Asian peoples.

over a very long period—in some cases as much as 1.5 million years—and in dissimilar places went through quite separate incarnations: Neanderthals, Peking Man, Java Man, and so on. And yet, over the same long time, there was enough—just enough—intermingling of these peoples, or "gene flow," everywhere to keep these otherwise separate populations evolving genetically together as one species, *Homo sapiens*.[22] In other words, there was just enough constant "shuffling" of the genetic cards to distribute the essential human species' genes evenly, while leaving some cards—those that determine racial characteristics—firmly static through the shuffling.

According to this scheme, *Homo erectus* of Java and Ngandong evolved into Australian Aborigines, the Peking and Dali fossils are the remains of people who formed the Mongoloid line, the Broken Hill skull, from northern Rhodesia (now Zambia) was part of a proto-Negroid or proto-Capoid population, and the Neanderthals became the Caucasoids. But does the fossil record really show these local sequences of change? Was there truly a global ladder of human progress with racial features formed early and maintained all the way along? Or, to modify our analogy of the constant shuffling of a pack of cards to represent multiregionalism, was human evolution instead more like a series of games of cards? Were particular hands dealt out to form the regional human populations of Europe, Java, China, and the rest, and were the cards sometimes restacked, producing extinction, followed by the deal of a completely new hand of cards, causing replacement? Perhaps the populations of Solo, Dali, and the Neanderthals represented hands that were restacked, rather than reshuffled—so that *Homo sapiens* can be viewed as the latest deal in the game of human evolution.

This idea forms an alternative view to multiregionalism and, in various guises, was put forward by several scientists earlier this century—such as Henri Vallois, a student, and then colleague, of Boule in France. He believed Europe's fossil record showed that one line of evolution in Europe had split off far back in the Pleistocene, and by 250,000 years ago, was already evolving towards modern *Homo sapiens*, while another line of more primitive fossils was evolving from *Homo erectus* to the Neanderthals—and ultimately to extinction.[23]

However, the American anthropologist Loring Brace, from the University of Michigan—in a vicious but highly influential attack on the "old guard" of human evolution—claimed people like Vallois had expelled the Neanderthals from our ancestry without coming up with any plausible alternatives:

> Recently many physical anthropologists have been clinging to the old view of a sudden migration into Europe of Upper Palaeolithic [anatomically modern] peoples, although they have been unconvinced by the skeletal evidence. According to them the proof is mainly archaeological. On the other hand, archaeologists have continued paying lip service to the sudden migration view with the feeling that the justification was largely based upon the supposedly clear-cut morphological distinctions made by the physical anthropologists.[24]

He argued instead for a worldwide Neanderthal stage in human evolution, one that was transformed more or less simultaneously around the globe. (In some ways, his scheme echoed that of Weidenreich and Coon, but while they recognized fundamental differences between local evolutionary lines and were convinced of their racial nature, Brace entirely rejected the concept of race.) For him, ancient fossils—once they had been properly investigated—would show a gradual and progressive development of Neanderthal-like features, while early modern humans would everywhere show signs of their immediate Neanderthal ancestry. Modern-looking people could not have evolved at the same time as Neanderthals because such peoples evolved *from* Neanderthals. Thus Brace challenged each and every ancient date claimed for an early modern human fossil, whether it was in Europe, Israel, Borneo, or Africa. His views were part of a general rethink which, in tune with the liberal sentiments of the 1960s and 1970s, began to treat the Neanderthals like a wrongly persecuted minority group. This was the time of Neanderthal Liberation as the pendulum swung back to restore their position as our rightful ancestors.

We can get a sense of this intellectual rehabilitation from William Golding's *The Inheritors*,[25] a book that contrasts markedly with those of Rosny-Aines and Wells in presenting a highly sympathetic picture of the Neanderthals. Indeed, Golding begins his book with a quote from Wells which stresses once again the "extreme hairiness, an ugliness, or a repulsive strangeness" of the Neanderthal. Golding then proceeds to depict Neanderthals in terms of noble simplicity, while the book's modern human protagonists, even though ultimately triumphant, are portrayed in distinctly unfavorable terms. Certainly the descriptions of Neanderthals are far less brutal than those quoted at the beginning of this chapter.

> The mouth was wide and soft and above the curls of the upper lip the great nostrils flared like wings. There was no bridge to the nose and the moon-shadow of the jutting brow lay just above the tip. The shadows lay most darkly in the caverns above its cheeks and the eyes were invisible in them. Above this again, the brow was a straight line fledged with hair; and above this there was nothing.

This restoration of Neanderthal reputations was taken a stage farther by a young American archeologist, Ralph Solecki, who began excavating at the Shanidar cave above the Greater Zab River in Iraq. The cavern was still intermittently inhabited by Kurdish tribesman who had even built small rooms round its edges and a large pen for their goats and horses—so Solecki dug his trench in the cave's central common area. Associations with this domestic clutter, together with the fact that Shanidar appeared to have been more or less continuously occupied for the past 100,000 years, greatly influenced Solecki, who became convinced he was investigating a seamless chain of events in the evolution of modern humans.

Between 1953 and 1960, Solecki found a total of nine Neanderthal skeletons at Shanidar.[26] Some may have been killed in rockfalls, though others appeared to have been intentionally buried, he thought. In addition, some of the bodies displayed dramatic, but old injuries—in one case, a fractured eye socket that had probably caused partial blindness, as well as a withered right arm, and damaged right foot and leg. These traumas did not result in the man's death, however. The victim lived on, his survival indicating that Neanderthals were capable of compassion and tenderness towards the sick, said Solecki. Certainly, this did not appear to be the behavior of brutes.

But the most dramatic of Solecki's discoveries did not involve bones or stones—but earth. Soil samples taken from one of the burial sites were found to be extraordinarily rich in pollen, much more than could ever have been blown in by the wind or carried on animals' feet. The inference was clear, said Solecki. The dead individual—an elderly male—had been buried with offerings of flowers. "The death had occurred approximately 60,000 years ago . . . yet the evidence of flowers in the grave brings Neanderthals closer to us in spirit than we have ever before suspected," Solecki wrote later. "The association of flowers with Neanderthals adds a whole new dimension to our knowledge of his humanness, indicating that he had 'soul.' " The fact that Solecki thought that Neanderthals had soul (without the indefinite article) and that he chose the title *Shanidar—The First Flower People*,[27] for his book now seems slightly risible with two decades of cynical hindsight, particularly as some archeologists today even question the evidence for those floral burials. Nevertheless, Solecki's work shows how powerfully opinion had shifted about these vanished people—for at the time no one demurred about its poignant vision of their compassionate, gentle nature.

However, the apotheosis of Neanderthal veneration arrived with the publication of George Constable's 1973 book, *The Neanderthals*.[28] For a start, it carried an introduction by Solecki, who argued that in the Neanderthal we see "the mind of modern man locked into

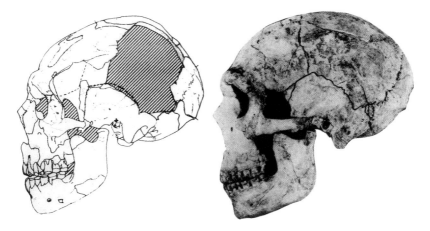

28   The Shanidar 1 skull (left) and Amud 1.

the body of an archaic creature." Constable's first chapter then proceeds with an argument that "there is much new evidence to demonstrate that some Neanderthals—and perhaps all of them—were our
immediate ancestors; they carried the torch of evolution during the
millennia from 100,000 years ago to about 40,000 years ago. . . .
Clearly, they were ancestors to be proud of." And later on, he states:
"Middle East fossils therefore serve to establish a solid evolutionary
link between Neanderthals and modern man." As for the doubters,
they are dismissed in the following terms:

> All those who relegated Neanderthals to a side branch of
> human evolution believed (and some still do) that modern men
> existed somewhere on earth during the Neanderthal era. . . .
> But if modern man existed so long ago, where was he hiding?
> Generations of scholars have devoted their careers to a search
> for very ancient but modern-looking ancestors . . . but each
> time the fossils . . . failed to fulfill their promise.

By the end of the book, Constable explodes in an eruption of
lyrical praise for these lost creatures:

> Place him in a landscape of tall, waving grass, with the sun
> shining down and the bubbling music of summer in the air.

Who is this man? He is an evolutionary bridge, just shy of fully modern status. He is a true human—our ancestor. We should regard him with honour, because almost everything that we are springs directly from him.

Now there is much to admire in the attitude of Golding, Solecki, Constable, and others. For one thing, it is refreshingly clear of those colonial attitudes which viewed foraging lifestyles as being inferior ones, and which have distorted and debased our vision of them in the past. These new views are commendably sympathetic, and tolerant of differences, if nothing else. But rehabilitating the Neanderthals is a different issue from demonstrating that they were the immediate ancestors of modern people. Indeed, by the mid-1970s the pendulum was beginning to swing again—away from the notion that these were evolutionary bridges that crossed the divide between ancient hominids and modern humans.[29]

Some of the first to walk down this road to scientific heresy were three researchers working in South Africa—Peter Beaumont, an archeologist excavating the Border Cave; Hertha de Villiers, who studied its fossils; and John Vogel, who did its dating. They had read Chris Stringer's first publication in the *Journal of Archaeological Science* and incorporated his results into their next paper in the *South African Journal of Science*.[30] The trio were part of a group led by the archeologist Desmond Clark, of the University of California, Berkeley, who believed science was looking in the wrong place for signs of modern humankind's origins. They were not to be found in Europe, but in Africa, Clark argued. He maintained the later parts of Africa's Stone Age had been wrongly dated. It was far older than generally believed. And if he and his group were right, Africa's Middle Paleolithic period occurred at about the same time as Europe's (between about 150,000 and 40,000 years ago). This suggested the continent was anything but a stagnant backwater in the tide of the affairs of modern mankind. For example, there were composite tools such as hafted spears and, in some sites, narrow blades of stone like the ones which dominated the technology of the Cro-Magnons. Indeed, there was some evidence to suggest that in Africa advanced

Stone Age culture might have developed much earlier—not later—than it did in Europe. At Border Cave, there were four early modern fossils—a partial skull, two lower jaws, and a tiny, buried infant—all, apparently, at least 90,000 years old. Other South African sites told a similar story. At caves near the mouth of the Klasies River, Middle Paleolithic tools were found with fragmentary but modern-looking fossils, stratified in layers immediately over a beach that had been deposited about 120,000 years ago. In other words, they had a sharp rejoinder to that question of Constable, who had asked, no doubt rhetorically: "If modern man existed so long ago, where was he hiding?" The answer is simple, they said. He was living in Africa.

The following year, the Frankfurt anthropologist Reiner Protsch used several different dating techniques, some established and some still experimental, to investigate a range of African fossils. He claimed Africa's early modern humans were ancestral to all later forms of *Homo sapiens*—though his paper was flawed in other respects.[31] Beaumont, de Villiers, and Vogel made an even more convincing case. They put forward the argument that a common population existed in Africa and Europe about 400,000 years ago (evolved from *Homo erectus* and which is often now called *Homo heidelbergensis*, replacing the term "archaic *sapiens*" used in early chapters, and which is represented by fossils like the Petralona and Broken Hill skulls, see illus. 22). Then there was an evolutionary split, a divergence nurtured by the growing geographical barrier of the arid Sahara desert. The lineage north of the Sahara became the Neanderthals of Europe and the Middle East, while the southern one became the first modern humans. The group argued that the game-rich savannas of southern Africa were the "satisfying and hospitable hearth" for a gradual emergence of modern humans, rather than the self-created cultural environment of the Neanderthals during the Middle Paleolithic of Europe or the Middle East, as Brace and Constable proposed.

The tide was beginning to turn. Very soon evidence, which would eventually drown multiregionalism, began to turn from a trickle into a flood, with some of the first rivulets appearing in the Levant. Here scientists had begun digging in 1929 when Dorothy Garrod, a Cambridge University archeologist, began a five-year excavation at the caves of Skhul, Tabun, and el-Wad near the mouth of the Nakhal HaMe'arot River in western Israel,[32] a program that was followed by subsequent digs by other archeologists at Amud and Qafzeh to the east, and at Kebara to the south. Among the many important finds that were made there was the uncovering of the remains of a woman, apparently deliberately buried, at Tabun. In addition, many other human bones were unearthed at the nearby site of Skhul.

Some of this material looked typically Neanderthal—large brow-ridges, thick-walled leg bones, and all the species' other distinctive appurtenances—while other remains were more gracile, and characteristic of modern humans. Treating this entire assemblage from these caves as a unit, researchers had concluded the skeletons represented a form of hominid intermediate between Neanderthals and modern humans. Those at Tabun and Amud seemed to be more ancient and more Neanderthal-like, it was decided, and were estimated as being about 50,000 to 60,000 years old. Those from Skhul

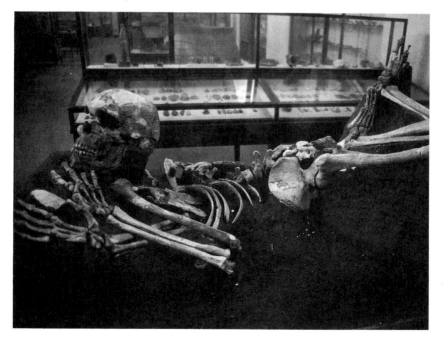

29   The early modern burial of Skhul 4 from Israel.

and Qafzeh were earmarked as being more modern and more like *Homo sapiens*. They were dated as being about 40,000 years old. In other words, these jumbled fragments were supposed to provide mute testament to the gradual progression of ancient Neanderthals into more gracile *Homo sapiens* in the Levant, confirming how humankind had evolved from our evolutionary cousins approximately 45,000 years ago.[33]

But by the 1970s, scientists began to realize the layers of sediments in which they had found these "protomodern" remnants were much more complex than previously appreciated. Unfortunately, they lacked instruments that could peer far enough into the past, with sufficient precision, to date the strata and their fossil contents. The only effective technology that was then available relied on radiocarbon dating, a technique that is only useful with remains less than 40,000 years old—which put the Levant sediments tantalizingly out of reach.

However, by the 1980s, new dating methods—which we shall meet in Chapter 6—were developed, giving scientists a startlingly new acuity when probing the past. Some exploited different forms of radioactive decay—involving uranium and thorium atoms, as well as thermoluminescence, which measures the effects of naturally occurring radiation in samples that had been burned (such as flints that have dropped in a fire), and electron spin resonance, which achieves the same feat using crystals such as those found in tooth enamel. These technologies permitted scientists to determine how old were the Levant hominid bones, and produce results that generated one of the greatest upsets in modern paleoanthropology. Yes, the Neanderthal remains—such as those at Kebara—were confirmed as being about 60,000 years old, while others were dated at around 40,000 to 50,000 years ago. So far so good. However, when flints and animal teeth found with the remains of those more modern-looking people from Qafzeh and Skhul were tested, scientists found they were about 100,000 years old—60,000 years older than the previous estimate and 40 millennia older than the Neanderthal remains from Kebara.[34/35] Yet the Kebara Neanderthals were supposed to be the ancestors of the Qafzeh humans! The arithmetic of mankind's recent evolution had been turned on its head. Neanderthals, far from being our evolutionary fathers and mothers, looked more like paleontological cousins—and rather recently arrived ones at that.

"No longer could the robust skeletons be identified as the ancestors of the gracile skeletons," stated archeologist Ofer Bar-Yosef and paleoanthropologist Bernard Vandermeersch, in a feature, "Modern Humans in the Levant," in *Scientific American*.[36] The ancestor-descendant relationship between Neanderthals and modern humans, that the multiregionalists so vehemently placed their faith in, suddenly looked wafer thin.

But if modern humans such as Cro-Magnons did not evolve from Neanderthals, where did they come from? Where do the roots of modern Europeans' immediate ancestors lie if not in Europe? One important clue came from Erik Trinkaus's research on Neanderthal and early modern skeletons. As we saw earlier in the chapter, local populations today have physiques suited to their places of origin, such as the Masai of Kenya and natives of Central America, who tend to have narrow cylindrical physiques, and large surface areas of skin from which to lose heat. Now this adaptation is revealed through greater relative length of body extremities, especially the shinbones of the lower leg when compared with the upper leg, and the bones of the forearm when compared with the upper arm. The opposite proportions are found in cold-adapted peoples such as the Lapps and Eskimos. In other words, the proportions of the leg and arm bones can act as a kind of limb thermometer, roughly indicating the average temperature of the land in which a population originated. Moreover, these ratios are predominantly determined genetically. African-Americans have mainly retained their African physiques, for example, while Afrikaaners' proportions still resemble those of their Dutch ancestors.

Trinkaus assembled data which demonstrated this clear relationship in human skeletons from around the world. Then he "plugged" the Neanderthals into his limb thermometer, and discovered the species had a shinbone that is, on average, only about 80 percent of the length of its thighbone. From this Trinkaus calculated an average predicted temperature of about 0 degrees Celsius (i.e., freezing point) for their habitat. This fitted the known Ice Age environments of Europe perfectly, and also suggested that Middle Eastern Neanderthals were showing signs of a European "cold" physique, as well. However, the real revelation came when Erik inserted his data on the Cro-Magnons of Europe and the Skhul-Qafzeh skeletons from Israel into the equations. In this case, he got a figure of 85 percent for the shinbone-thighbone ratio. Not only were they unlike the Neanderthals, but these people actually fell at the other extreme in their readings on the limb thermometer. The predicted average temperature of origin for folk with an 85 percent shin-thigh fraction—indicating much longer extremities relative to trunk length—was about 20 degrees higher than the Neanderthals', suggesting a subtropical—if not tropical—homeland![37]

Similar distinctions in face shape were also revealed by Chris Stringer's studies. As we saw earlier, multiregionalists claim there is a fundamental similarity between Neanderthals and modern Europeans in their large, prominent noses. So Chris Stringer decided to compare this feature in Neanderthals, the earliest Cro-Magnons, and recent Europeans. Given that the nose (which is made of flesh and gristle) is an anatomical feature that conspicuously does not survive the passage of time, that might seem tricky. There are ways round the problem, however. One can get a reasonable indication of nose size from a fossil by measuring the distance between the base of a skull's nasal opening, and the nasion, the point between the eyes. You can also measure the width of the nasal cavity, and how much it protrudes from the side of the face. When this information was run through a computer, it was found that while Neanderthals and modern Europeans are, indeed, somewhat similar, Cro-Magnons—the supposed descendants of Neanderthals, according to the multi-regional hypothesis—again stood out as different. They had relatively smaller, flatter noses than either the Neanderthals or modern Europeans.[38] It seems they came into Europe looking different, and then evolved a nose that ended up looking more like a Neanderthal one. Given time, the European environment was apparently shaping an intrusive warm-adapted people (the early Cro-Magnons) into a population which in some ways became more like the Neanderthals in face and body shape—a process called parallel evolution.

Then there was the evidence we encountered in the first chapter, which was provided by those slender, elegant bones of the Kibish man, a modern human being who lived in Africa more than 100,000 years ago. In addition, the archeological studies of Yellen and Brooks suggested the relative sophistication of life on that continent at that time. The results were becoming very convincing, although not every one of these new African apostles was prepared to write off the Neanderthals completely as potential ancestors. Günter Bräuer, of Hamburg University, was also a strong early advocate of the Out of Africa theory, but he still believes that some interbreeding between Neanderthals and early Cro-Magnons occurred in the Middle East and Europe, as demonstrated by the mixed features of the Hahnöfersand skull bone from Germany, and other fossils.[39] Other workers were receptive to the new African data, but still maintained Europe was an important arena in modern human evolution. For example, Fred Smith, of Northern Illinois University, argued that "in central Europe we can demonstrate that we have late Neanderthals that are to a very great extent good intermediates between earlier, more primitive Neanderthals and early modern central Europeans."[40] Most forceful of all, however, were the words of Milford Wolpoff. In typically abrasive style, he denounced the Out of Africa theory as the Killer African hypothesis:

> The spread of humankind and its differentiation into distinct geographic groups that persisted through long periods of time, with evidence of long-lasting contact and cooperation, in many ways is a more satisfying interpretation of human prehistory than a scientific rendering of the story of Cain, based on one population quickly, and completely, and most likely violently, replacing all others. This rendering of modern population dispersals is a story of "making war and not love," and if true its implications are not pleasant.[41]

At no point, of course, had any Out of Africa proponent suggested a violent replacement of Neanderthals by Cro-Magnons had to have taken place. As Chris Stringer had stated in 1984, "evolutionary events in Africa may have led to the emergence of the Cro-Magnons, whose intrusion into Europe seems to have led to the demise of the Neanderthals—not overnight, but after a period of coexistence of several thousand years."[42] He argued that the two species may have coexisted peacefully for several thousand years before the former died out because it could not compete, economically, with the latter (an idea we shall explore in much greater detail in the next chapter). This point was ignored—but then attention to inconvenient details has never been part of the Wolpoff style of rhetoric.

As for Loring Brace, he wrote in 1994—in characteristically polemical style (i.e., open derision, mixed with a dash of xenophobia and a fair measure of sarcasm)—that "Stringer's basic stance is a splendid example of what has been termed the 'great leap backwards' that has characterised so much of palaeoanthropology over the past decade and more. One could say that the [anti-Darwinian] spirit of Sir Richard Owen is alive and well at the British Museum [sic] now a full century after his death."[43] (In fact, Chris works at the Natural History Museum.)

Such bluster was to no avail. By this time, the Out of Africa theory had already had such a strong impact that the equally mighty Museum of Natural History in Washington, D.C., had been forced in the wake of public protests to close part of its human evolution exhibit in 1991 because it did not reflect current thinking on the recent African origins of modern humans. The "great leap backward" appears to be as influential in Brace's home country of America as it is in Britain. Indeed, in the next two chapters we will show—for all the fury of Brace, Wolpoff, and others—that this "great leap backward," as they would term it, has been embraced with alacrity over the past few years by a growing body of scientists. Growing knowledge of the prehistory of modern humans and Neanderthals, and of the genetic evolution of *Homo sapiens*, has pushed the Out of Africa theory from a minority view to its present preeminent position in only ten years. The heresy of yesterday has become the radical truth of today.

We can therefore see a logical progression to our narrative. We have learned how science has pieced together the story of how an

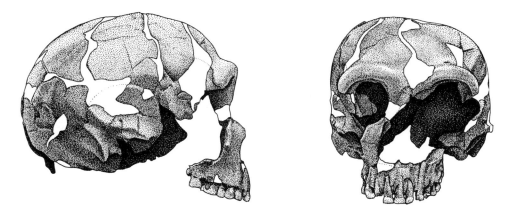

30　The skull of the Tabun Neanderthal burial from Israel.

upright, small-brained ape gave rise to several different hominid lines and eventually—after five million years of evolution—led to the emergence of *Homo sapiens*. We have also seen how one group of our immediate predecessors, the Neanderthals, were once viewed as grisly brutes but have slowly gained a place for themselves as an intelligent species in their own right—although, at the same time, we have learned that they are not the ancestors of human beings today, but are more like respected evolutionary siblings or even cousins. We shall examine further evidence which supports this point in succeeding chapters.

An obvious puzzle remains, however. If we did not evolve from Neanderthals, but were not much different from them, why did we replace them? There is no conspicuous evidence in the fossil record to show that we were stunningly cleverer than they were. Yet in the midst of the freezing wastes of Ice Age Europe we supplanted these solid, thick-boned people even though they were cold-adapted, and we were not. Finding answers to that mystery touches on the most fundamental aspects of being a member of *Homo sapiens*, an investigation which was only triggered with the realization of our recent African past. We shall now examine the enormous implications of this new understanding of human nature by comparing our own immediate ancestors with their Neanderthal cousins—although we must remember that the business of hominid replacement was not restricted to Europe but took place across the globe, in Asia,

Indonesia, possibly even Africa, and elsewhere, and involved the extinction of many more long-lived human lineages than just Neanderthals. However, we know so much more about Neanderthals and Cro-Magnons, because so much more fossil excavation and research has concentrated on Europe compared to other parts of the Old World. In any case, the lessons we learn from these extinct people can illuminate the fate of others—and tell us much about ourselves in the process.

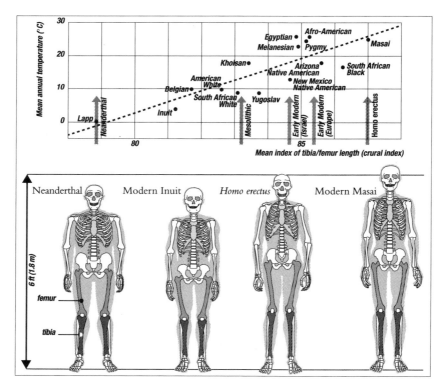

31 Modern values of the crural index (top) can be used to "predict" the temperature under which fossil humans evolved. Reconstructed skeletons suggest that African *Homo erectus* was adapted to an environment like that of today's Masai, while the Neanderthals showed an even stronger cold adaptation than the modern Inuit.

4

Time and Chance

The past is a foreign country; they do things differently there.

L. P. Hartley, The Go-Between

In 1992, in an old limestone cave, high above a riverbed in Upper Galilee, anthropologist Yoel Rak made a remarkable discovery. He uncovered the 50,000-year-old grave of a tiny infant. It was a poignant finding, a reminder, it might seem, of mankind's ageless urge to honor and treasure its dead. There was one flaw in this ancient, funereal interpretation, however. The bones within the "tomb" were not those of a human child—at least not one that we would ever expect to see in a crib or stroller.

The remains certainly resembled those of a member of *Homo sapiens*, but the jaw was chinless; the hole in the base of the skull, on to which the spinal column latched, was oval not circular; and the lower jaw was strangely structured. "This was not a member of our species," states Professor Rak with absolute certainty.[1]

In fact, Rak, of Tel Aviv University, had found the splendidly preserved remains of a ten-month-old Neanderthal child that had lain for fifty millennia inside the cave of Amud,[2] a vaulted cavern that takes its name from the distinctive rock column that stands like a sentinel at its mouth. (Amud is the Hebrew word for pillar.)

It was one of the most exciting moments of Professor Rak's career, and represented the culmination of a lengthy association between scientist and site, a relationship that can be traced back to 1959 when the cave was first excavated by Hisashi Suzuki of Tokyo University. Frustrated by postwar refusals to be allowed to dig in other countries, the great Japanese paleontologist had come to Israel and chose—for reasons that are now forgotten but which were nevertheless inspired—to excavate at Amud. He and his team began to dig a meter-wide trench across the cave and, after two years' work, eventually unearthed the shattered skeleton of an adult Neanderthal.[3]

When the news was announced, the young Yoel Rak, already fired with an enthusiasm for all things fossil, walked out of school and hitchhiked to the area. Unannounced he scrambled up to the cave, stumbling on a bizarre tableau of Japanese formality: the distinguished scientist sat under an umbrella, dining alone on one side, while his staff ate in silence nearby. The youthful interloper was treated with civility, however, and his nascent love affair with paleontology was consummated.

After Suzuki departed (taking the bones and stone tools that he had found back to Japan for study) the cave was abandoned until 1991, when Rak—now a professor of anatomy—returned, to help run a dig there along with archeologist Erela Hovers, of the Hebrew University of Jerusalem, and William Kimbel of the Institute of Human Origins in Berkeley, California. In the baking Israeli summer heat the team began to sift again through the dust and earth. They found little of note, but returned the next year, armed, as usual, with a posse of enthusiastic young students, those foot soldiers of human paleontology who toil, often in stunning heat, from 6 A.M. to dusk, using scalpels and brushes to scour the dirt and soil for signs of human bones or artifacts.

For weeks they collected nothing but stone flakes and animal bones from the cubes of earth that they had marked with string and flags, like giant three-dimensional crosswords. Such fragments, their positions carefully noted and stored in computers, help researchers to understand how, and when, a cave has been used. For example, by studying pieces of stone tool, it can be determined if these implements were fashioned on location, or, if there are no slivers leftover from knapping, whether they were made elsewhere. This, in turn, gives an indication of whether a cave was used merely as a resting place, as temporary quarters, or if it was a long-term accommodation center. Similarly, bones of certain rats, mice, and other vermin—those species which are our unwelcome fellow travelers—provide data about the permanence of the residence. The more bones of those particular rodents which live off human refuse, the more intense had been the occupation of the site. It is important work but not the stuff that makes reputations.

Then a student, scraping at a patch of ground near the cave's north wall, exposed the dusty outline of a distinctive piece of bone. It was the moment every paleontologist dreams of—the first faint glimmering of a big find. "We started brushing away the dirt, slowly, slowly. And there it was: a skull. It started with a skull. A very tiny skull, the size of my fist, started emerging. It was a very exciting thing," recalls Rak.

Gradually, the team exposed the skeleton of the infant and its resting place: a rock niche below an alkaline limestone shelf that had shielded its skeleton from eons of pounding by feet and hooves, and from destructive accumulations of acidic guano and animal urine. It was this fortunate geological accident that had protected the bones of the little body, leaving them—even after 50,000 years—in an almost complete state, in contrast to most of the frustratingly skimpy remains of early hominids that are usually unearthed by paleontologists.

Indeed, the evidence suggested it was not just chance that had protected the child's remains. There seemed to be signs of purpose behind its interment. In other words, it appeared that the body had been deliberately placed in the ground, and covered over. For one thing, the skeleton was articulated: its form was relatively intact, and its bones laid out in position with its arms pressed by its side. This completeness suggests the burial must have been a fairly prompt business, otherwise scavenging hyenas or wolves would have stolen and scattered its bones.

Of course, a cave roof-fall could have done the trick. Or perhaps the body had only been covered up to eradicate the stench of decay. However, the team was to find even more striking evidence that they had uncovered the remains of a primitive funeral, for as the excavation proceeded, their careful scraping revealed the jawbone of a red deer which had been placed over the pelvis of the reclining infant. Rak interprets this as a grieving, parental tribute. "It was an intentional offering," he says, "though it is not clear whether it was meant as food for the afterlife, or as some more symbolic gesture. However, this was a primary grave, and the baby was deliberately put in it."

Rak's discovery has an eerie, touching quality to it: another species possibly caught in an act of veneration to an afterlife, a performance tinged with irony given the eventual demise of the entire line. But if the burial had the unhappy imprint of human bereavement, Rak's other investigations indicated that the Amud child's relationship with *Homo sapiens* was still a relatively remote one, more like that of a cousin than an immediate evolutionary sibling. In particular there was the arresting, alien nature of the child's anatomy, its physical attributes distinguishing it instantly from our own species.

"It is hard to tell the difference between the skeletons of any infant primates—chimps from gorillas, for instance—and they are completely separate species. When they are very young, their anatomies are all very alike. But this was instantly recognisable. It was just not human in a way that we would recognise," Rak observed. And that has important implications, for if Neanderthals looked so alien when very young, just think how different they must have appeared as adults. We are dealing with a completely distinct species compared with modern humans.

In anatomical terms, the three singular features that made Rak so sure of his diagnosis are the absence of a mental protuberance (i.e., a chin) on the front of the lower jaw; the highly elongated shape of the foramen magnum (the opening of the skull base through which the spinal cord passes); and marks which showed the presence of what is called a medial pterygoid tubercle, an attachment point for the stylo-mandibular ligament which helps to hold the jaw in place and which is strongly developed among Neanderthals. In this latter case, it was already known that the species possessed a particularly large ligament, indicating its members were capable of very powerful chewing, their jaws acting like third hands, so clamplike were their powers. Plants, roots, nuts, and flesh would have been ground between Neanderthal teeth like automobiles in a car crusher. We, by comparison, are—by and large—wimps when it comes to the business of chewing.

But did Neanderthals develop powerful tubercles and prowess at masticating merely because they did a lot of chomping on any vegetation that came to hand—and so built up strong jaws? In such a scenario, muscle development would have been an environmental response to a lifestyle challenge. On the other hand, was it equally possible that Neanderthals were born with this intrinsic difference? Well, the skull of the Amud baby provided the answer, says Rak. Even ten-month-old Neanderthals had well-developed tubercles. In other words, they were genetically programed from conception to be powerful chewers. It is another sure sign, adds the Israeli anthropologist, that Neanderthals were inherently different from modern humans.

It is an intriguing picture—and a seemingly contradictory one. On one hand, there is evidence of complex behavior and sophisticated, symbolic thought processes that are manifested today only by *Homo sapiens*: in this case, an awareness of mortality and possibly even a belief in an afterlife. By contrast, there is definite evidence of a uniquely specialized anatomy, a sure sign of divergence from the evolutionary line that later led to modern human beings. Nor are these latter differences restricted to the three uncovered by Rak. We also know that Neanderthals were bulky and squat with heavy muscles and barrel chests compared to *Homo sapiens*. Males were, on average, 5 ft. 6 in. tall and weighed about 140 lb., while females were about 5 ft. 3 in. and weighed about 110 lb. This basic form varied according to locality, however. The sturdiest and stockiest populations lived in northwest Europe, a colder region near the ice sheets that then shrouded the higher latitudes, while somewhat more lightly built Neanderthals lived in east Europe and west Asia.

This was a physique adapted to a harsh existence, powerful muscles suggesting a more robust, physical response to the rigors of survival. "These were creatures with physical prowess far beyond the aspirations of even the best Olympic athletes," says John Shea of Harvard University.[4] Indeed, all the evidence indicates that they lived "a hard, violent life," as Shea puts it. "There is the near ubiquity among Neanderthals of healed head, arm, and leg traumas. In addition, age estimates suggest that few if any Neanderthals lived beyond their thirties and forties to a postreproductive age." Equally importantly, their frames imply that Neanderthals' bodies were shaped between 100,000 and 300,000 years ago by repeated periods of glaciated chill that then gripped Europe. As we have seen, they had a round, thickset build, while early *Homo sapiens* were lankier and less compact.

On the other hand, modern humans and Neanderthals also shared many features: large brains; upright stances; long childhoods; meat-eating; fire-making; shelter and stone tool constructing; and an ability to speak. However, in every case (except the first), these similarities need qualification, for each depends on interpretations of fossil or archeological evidence. Nevertheless, a look at this list of common attributes sheds an intriguing light on the Neanderthals—and ourselves. The former were an evolutionary dead end. So far, we have avoided this fate, which means that somewhere in the anatomical and archeological evidence there may be clues to those crucial attributes whose absence doomed the Neanderthals and which aided us so spectacularly.

So let us move from Neanderthal head to toe, starting with the brain. Its volume ranged from about 1,200 to 1,750 ml. in adults (the biggest ever found being the adult Neanderthal that Hisashi Suzuki discovered at Amud). Modern humans show a similarly large range, but with a lower average size, with females being slightly smaller than males in both cases (though the differential is slightly less pronounced in *Homo sapiens*). However, the Neanderthal brain was flatter on top, smaller at the front, and bulged more at the sides and back. Some researchers believe these distinctive contours indicate that our biological cousins had larger visual cortexes and smaller frontal lobes, a brain area which is associated, in modern people, with planning actions. In other words Neanderthals may have been better observers than strategists.

We should not make too much of the idea, however, for it is notoriously difficult to assess brain quality and purpose from its surface features. Cerebral function depends on the complexity of its internal wiring—though researchers do agree on one issue. Neanderthals' brains were asymmetrical: slightly larger at the right frontal and left rear, indicating—as it does with modern humans, and as it did with *Homo erectus*—that they were right-handed. In other words, the species was dextrous at least in one sense of the word.

Then there was that glowering, twin-arched Neanderthal brow-ridge, itself slightly smaller than its *Homo erectus* predecessor (though it still dwarfed *Homo sapiens'* virtually nonexistent browridge and accompanying high forehead). Why our ancestors evolved this anatomical curiosity two million years ago remains a mystery. Possibly it was used as a signaling or threat device (mainly between males)[5] or to protect the eye sockets, which would otherwise have been exposed by their lack of forehead. Or possibly their function was even more utilitarian and prosaic, as Grover Krantz, of Washington State University, demonstrated in an experiment of delicious eccentricity. He made a replica of a *Homo erectus* browridge, strapped it above his eyes, and wore it for six months. Thus Krantz was rendered a fairly startling sight on a dark night (supporting the signaling hypothesis, if nothing else), while during the day his hominid headgear kept the sun out of his eyes. However, Krantz found the browridge's greatest success lay in keeping his hair out of his eyes, since his flowing locks (which he also grew as part of his experiment) parted neatly over either side of the browridge! "Several other long-haired people have tried out these

32   Grover Krantz wearing his *Homo erectus* browridge.

artificial browridges with the same result," adds Krantz (a man with many like-minded friends, it would seem).[6] In other words, *Homo erectus* evolved beetlebrows to stop their locks flopping in front of their eyes, and then passed this advantage on to their Neanderthal successors, for as Krantz puts it, "similar conditions evidently continued through the so-called Neanderthal stage of human evolution."

Neanderthal eyes were presumably large, judging by their big, round, skull sockets. However, we have no idea what color they were. African apes and most modern tropical peoples have brown eyes, suggesting a longer pedigree for this color. Equally, Neanderthals could have evolved blue eyes (along with paler skin), as did some modern humans as they adapted to Europe's poor sunlight.

However, the most remarkable Neanderthal organ was—as far we can tell from fossil evidence—the nose. These must have been phenomenal objects, dominating their faces like anatomical flagpoles. Beaky noses are common enough today (among Europeans and native Americans, for example) as are broad noses (among natives of Africa and Australia), but the Neanderthal nose combined both projection and width. It stuck out horizontally between the eyes, and its prominence was accentuated by cheekbones that were swept back on either side. Inside, the floor of the nasal opening was lowered (compared to all other humans), providing its owner with enormous, corridor-like nostrils that would have had very large surface areas of moist skin and good blood supply. And on either side of the nose, there were huge spaces, within the cheekbones, called sinuses, that could have stored warm air for insulation. Modern humans have them, but Neanderthals had positively cavernous ones. "You could put a swimming pool in those sinuses," says anthropologist Jeffrey Laitman, of the Mount Sinai School of Medicine.[7]

But what evolutionary advantage could there have been in having a protuberance like the tower of Lebanon? A heightened sense of smell, perhaps? Well, possibly, though such olfactory powers would have bucked the general trend of primate evolution which has de-emphasized the sense of smell, in favor of sight, over the past 40 million years. In any case, most mammals with such a sense usually have flat noses with moist tips, like dogs—but unlike Neanderthals.

Instead, most ideas about Neanderthal nose size concentrate on climate. For example, some scientists believe the whole middle face was pulled forward to distance it from the brain. Cold air could then be warmed up before reaching the vicinity of the cortex, which would have required a very stable temperature and blood supply. Lots of warm moist air would have been puffed out from the lungs, because of their energetic lifestyle, to be mixed with cold, dry air being breathed in. Their spacious nasal cavities would have made excellent heat and moisture exchangers for improving the quality of inhaled air before it reached the lungs.[8]

Now climate might explain the internal volume of the Neanderthal nose but does it also account for its projection? Why should it have stuck out so far, particularly when it must have got chilled badly in the freezing weather that was typical of the time? This is an awkward question to answer, but its towering presence could also have had a nonbiological purpose, and may have operated as another strong visual signal. In other words, the process of sexual selection, which we encountered in Chapter 3, might also have been partly responsible for the projectile Neanderthal hooter. Mate preferences— related to alleged "beauty" or "handsomeness"—may have focused, rather bizarrely, on nose size. In other words, Neanderthal men and women may have picked their mates for their big noses, while natural selection may have favored a large internal volume. Both factors could have been involved.

As we have seen, the Neanderthal jaw had powerful muscles and ligaments, particularly those involved in clamping it shut. The front teeth were large—an odd feature because humans have evolved relatively small front teeth (incisors) compared with apes. In addition, Neanderthal teeth were anchored by long roots and show signs of heavy use, those at the front being ground down as though something had repeatedly been pulled through tightly clamped jaws. Perhaps Neanderthals were stripping plant stems or softening hides by pulling sections through their teeth, as Eskimos do today. On top of this dental stigmata, scratches on many teeth indicate where something— perhaps meat—was held in the jaws and sliced by stone tools, with direction marks indicating that these individuals were right-handed (as we also saw with the Atapuerca people that we described in Chapter 2).

Trying to decipher lifestyle from fossilized bones is clearly an intriguing, though rather complex business. These problems of interpretation fade into insignificance, however, when we come to the next set of features—those concerned with speech, perhaps the most vexed and hotly debated of all topics concerning Neanderthal anatomy. As we shall see, major changes in the position of the larynx—that bundle of cartilages, ligaments, and membranes that guards the opening into the windpipe, and allows us to produce the full lexicon of spoken language—began to evolve in hominids from the *Homo erectus* stage onwards. The base of the skull began to shorten, allowing the larynx to drop, and with a larger air space above it, a far greater range of sounds could be created. Mankind had set off down the road to the "perfect plainness of speech," and one can see a steady sequence of deepening larynxes progressing from australopithecines to modern humans. "Now you have a big air space up above the larynx that can modify sounds to a greater extent than in any other mammal," says Jeffrey Laitman in an interview in *Discover* magazine.[9] "What you've done, essentially, is taken a bugle and turned it into a trumpet by adding more tubing. This is what gives us the physical ability to produce fully articulate speech. It's distinctly human." There is one exception to this oral advancement, however—the Neanderthals. "The feature has actually gone in a reverse direction, away from this steamroller toward increased vocal communication in our lineage. That's very hard to explain," adds Laitman. In other words, in Neanderthals, the skull base seems flatter and the larynx could therefore have been higher than in the hominids that came immediately before them.

To many anatomists and paleontologists, such an observation fits with their views of the Neanderthal as a "derived" species, a specialist, one-off diversion from the main hominid line that evolved from *Homo erectus* to *Homo sapiens*. For example, while *erectus* and *sapiens* both have flat or hollowed cheekbones on the fronts of their skulls, in Neanderthals they are inflated and pulled forward to become almost vertical to the plane of the face. The retrograde direction of their larynx is another example. "These were very specialised creatures," says Rak. "They were an outgroup, a dead-end."[10]

The reasons for the Neanderthals' apparent vocal backsliding may be quite straightforward, however. Perhaps a return to an elevated larynx would have constricted the area behind the mouth, making it harder to gulp in air. And that would have meant taking in smaller mouthfuls of that freezing European atmosphere that would have otherwise wreaked damage on their throats' and lungs' delicate membranes. Instead, most air would have been inhaled through those mighty nasal heat and moisture exchangers, so protecting internal tissue. Nor was the Neanderthal larynx necessarily a dead loss. It may not have been able to articulate some vowel sounds, like a, i, and u, as well as we can, but that would not have prevented Neanderthals from talking to each other. Many modern languages neglect the full range of vowel and consonant sounds that can be made by the human throat, without limiting effective communication.

The base of some Neanderthal skulls was certainly distinctively shaped. In modern humans, it is folded or "flexed," but in some Neanderthals it has an open, flatter cranial underpinning. And since this base anchors the structure of the throat to the skull, it is argued that this smoother surface also indicates the limited vocal range of the species. On the other hand, the discovery of a hyoid bone—which is associated in *Homo sapiens* with the business of articulation—in a 60,000-year-old Neanderthal (from Kebara) suggests to other researchers that Neanderthals did have a relatively modern vocal tract.[11]

At the end of the day, if we want to know whether the Neanderthals had language of a modern human type, we really need to learn a lot more about their brains. And as we move away from the simple, almost phrenological, notion of trying to record the presence of speech by reading the bumps on brain casts or impressions on the insides of Neanderthal braincases to the realization that the successful production of human language involves many parts of the brain acting in concert and in sequence, we also move away from any realistic chance of directly assessing the quality of the Neanderthal brain, apart from the vaguest generalizations.

Moving below the neck, things become clearer, for we have now uncovered enough Neanderthal fossils to get a good impression of their bodies which, as we have said, were also more compact, larger, wider, and more barrel-chested, with shorter limb extremities. Neanderthal limb bones also have greater curvature than those of modern people, and together with their greater bone thickness, this suggests there were considerable stresses and strains during everyday life. The hands were not apelike, but were nevertheless built for a powerful grip, with deep muscle scars and enlarged fingertips. For these people, brawn was as important as brain in solving problems, and even Neanderthal women and children would have astonished us with their strength and power. The remarkable thing is, of course, that the Neanderthals were really only retaining a bodily pattern which had dominated the previous two million years of human evolution. It is us, modern humans, who are the exception. We have broken away from the demanding

33 Reconstruction of a Neanderthal "family" in Gibraltar by the late Maurice Wilson. While the physique is probably accurate, skin and hair form are entirely conjectural.

ancestral straitjacket of bodily strength and have opted instead for the life of a hominid wimp.

And lastly we come to the hips. At one time it was thought that Neanderthal women had larger pelvic bowls and birth canals for the development and delivery of larger, more mature, babies. However, recent discoveries now indicate this is unlikely. The Neanderthal birth process, like that of modern humans, was probably a difficult one, involving the baby turning head down and twisting during the delivery. Although the Neanderthal pelvic girdle was about the same as ours in internal volume, it was, however, different in its depth. Flaring of the hipbones at the side of the body made them lie further apart at the front. "My guess is that the modern human pelvis represents the unique adaptation," says Rak. "It would be an energy adaptation to long distance walking or running."[12]

In short, Neanderthals betrayed surprisingly "humanlike" attributes, but equally demonstrated a form that shows they were not members of our species. When we look at their bones, we gaze upon a people different from our own. Of course, both Neanderthals and our own immediate predecessors, such as the Cro-Magnons, would both look very savage and wild to us, if suddenly dropped into a modern shopping mall. Nevertheless, some researchers have argued they were "basically just like us." For example, William Straus and A.J.E. Cave believed that if a Neanderthal "could be reincarnated and placed in a New York subway—provided that he were bathed, shaved, and dressed in modern clothing—it is doubtful whether he would attract any more attention than some of its other denizens."[13]

Many other anthropologists dispute this point, however. They believe there would still have been an order of magnitude of difference between the appearance of a Neanderthal and an early *Homo sapiens*, as Steve Jones, professor of genetics at University College London's Galton Laboratory, has pointed out in a decidedly pithy riposte to Straus and Cave's subway scenario. "Most people would change seats if a Cro-Magnon sat next to them on a train," he says. "They would change trains if a Neanderthal did the same thing."[14]

These physical differences raise a broader issue, one that not only focuses a spotlight on our evolutionary cousins, but also casts an illuminating glow upon our own unique, innate sense of "human-ness"—for if we know what we are not, we will have a much clearer idea of what we are. As a species with a slightly different sort of intel-lect, physique, and behavior, Neanderthals provide us with clues in the most ancient of all searches—to know ourselves. And that, of course, is why they are so important in attempting to understand human nature. The species disappeared from the evolutionary scene 30,000 years ago, a mere 1,500 generations in the past, manifesting some modern human characteristics, but not all. So what were those crucial qualities possessed by us but lacking in them? What piece of anatomical or behavioral baggage might have weighed them down behind modern humans, leading to their extinction? Delineating these traits—which, for better or worse, gifted *Homo sapiens* with global supremacy—should tell us much about ourselves. Touching upon the very quintessence of being human, they constitute one of the most exciting tasks of modern science.

Unraveling our relationship with Neanderthals is clearly not an easy business. Science has shown they lived from at least 200,000 to 30,000 years ago, from Wales, in the northwest, to Gibraltar in the southwest; from near Moscow in the north, to Uzbekistan in the east. It is nevertheless a tight, geographical demarcation, for no trace of a Neanderthal has ever been found in any other part of the Old World—in Africa, India, or any part of eastern Asia. However, it is at the very limits of this terrain, particularly in the south, that some of the most intriguing, and important, aspects of Neanderthal existence have been unearthed, particularly those concerned with their rela-tionships with *Homo sapiens*, and especially those found in the Middle East. These finds include the discoveries at Skhul, Qafzeh, Kebara, and Tabun.

By inverting the timescale that had been established—wrongly— by those who saw Neanderthals as our ancestors, scientists have "shaken the traditional evolutionary scenario," as Bar-Yosef and Van-dermeersch have put it.[15] Far from being a static, geographical cru-cible in which Neanderthals gradually evolved into modern humans, the often parched terrain of the Levant has been revealed to be the dynamic setting for a strange pas de deux that was played out, 50,000 to 100,000 years ago, between these two ancient protagonists: thick-set, cold-adapted Neanderthals and that graceful upstart, *Homo sapiens*. These were not progenitor and progeny. They were contemporaneous settlers, and along the Mediterranean shore, and across parts of the Levant, these two hominid species indulged in an incessant bout of parry and thrust. Sometimes a valley was occupied by *Homo sapiens*. On other occasions it was the preserve of Neanderthals. The caves of Amud, Tabun, and Shanidar are associated with Neanderthal finds, for example, while those at Skhul and Qafzeh are linked with those of moderns.

But what were the Neanderthals doing in the relatively balmy Levant in the first place? Their presence may just have represented an excursion to the most southern end of their normal range. Alternatively, as Bar-Yosef and Vandermeersch point out:

> ... the Neanderthals were adapted to cold, as their compact bodies attest, but even they could not brave the Arctic conditions that occurred, in fairly sudden cold snaps, during the period between 115,000 and 65,000 years ago. The intense cold might have forced them southwards ... perhaps through modern-day Turkey or the Balkans.

And if they headed south, what did modern humans do? More suited to sunnier climes, they may well have found both a cooler Levant, and Neanderthals, just too much to bear and temporarily lost their foothold there, making room for the Neanderthals. Then during warm periods, they may have returned as these northern interlopers lost ground again. At times, the two populations may have held their ground and the two species would have had to coexist.

At present, the archeological evidence is inconclusive, and it is hard to tell just how much hobnobbing went on between the two, and how close was their contact. "It is impossible to tell what intermingling took place between ourselves and the Neanderthals," says Rak. "We could have missed each other by thousands of years. We do not have an accurate picture for population movements and habitations so long ago." Intuitively, though, it is unlikely that Neanderthals and *Homo sapiens* passed each other with very little contact, as Rak acknowledges:

> We have got to get rid of an idea, now deeply ingrained in our conscious, that because there is only one species of human being today, this has always been the case. For most of our evolution the opposite was probably true. There were at least two. Deeper in our past, there may even have been more. Think of that scene from *Star Wars*—in the bar where you see all kinds of aliens playing and drinking and talking together. I believe that image gives a better flavour of our evolutionary past.[16]

This vision of wildly different creatures sharing drinks, and indulging in the odd intergalactic bar fight, is rather outlandish when contemplating our cave-dwelling predecessors. Nevertheless, Rak's metaphor has merit, for it implies not just that several species could coexist relatively stably but that they could also share a common technological framework. In *Star Wars*, it was laser guns and spaceships. In the Levant, 100 millennia ago, it was the Middle Paleolithic stone kit.

And here we come across yet another complexity which makes a nonsense of any simple notion that one of these two species is the ancestor of the other. Certainly, Neanderthals showed possible signs of symbolic thought and ideas about the afterlife, but equally they displayed quite distinct anatomies from modern humans. And yes, the two species apparently lived separate lives, on a quite distinct timetable, in the Levant. But no, they did not use different tools while they lived these distinct lives. They appear to have used the same stone appliances. Yet, in one case, these implements helped bring long-term survival, as part of a developing new lifestyle, and in the other, they led their users along the road to nowhere.

Now this last point is a particular puzzle. If we assume there was a basic intellectual disparity which separated these two Paleolithic species and which accounts for the fact that one eventually thrived at the expense of the other, then why was this cerebral discrepancy not manifested in the most obvious way—in the utensils fashioned by the two species and around which they constructed their lives? After all, the hand is the cutting edge of the mind, as Jacob Bronowski pointed out.[17] Nevertheless, archeologists can find virtually no dissimilarity in types of tools used by Neanderthals or *Homo sapiens* in the Levant at the time. In both cases, they used Middle Paleolithic implements which consisted mainly of flint scrapers and knives, with occasional daggerlike points and hand axes sometimes being manufactured as well. Exactly how they were used by either species, we cannot say. All we can point to is the enigma of their universal use at that time and locality, contrasted with the dichotomy of their success with one species, and their "failure" with another, a point that is stressed by John Shea:

> If Neanderthals' abilities to use stone technology are taken as a measure of their intelligence and adaptive ability, then the archaeological record suggests few major differences between Neanderthals and early modern humans. In Neanderthals our ancestors would have confronted hominids perhaps very different in appearance, but every bit as intelligent as themselves. Why we are here today and they are not, is one of the most intriguing questions in palaeoanthropology.[18]

To some scientists, such as Yoel Rak, the problem is simple. It belongs to someone else. "Why should I worry why two distinct species use the same tools? I am an anatomist. I am telling you they are different creatures. Someone else must explain why they had the same kit. It is an archaeological problem."[19]

This is a fair point. Yet the difficulties in explaining why two separate species used the same tools has serious ramifications for the Out of Africa theory, for it presents its opponents with one of their best opportunities to question its implications. "These 'modern' people had a culture identical to their local Neanderthal contemporaries: they made the same types of stone tools with the same technologies and at the same frequencies; they had the same stylized burial customs, hunted the same game and even used the same butchering procedures," state Alan Thorne and Milford Wolpoff in *Scientific*

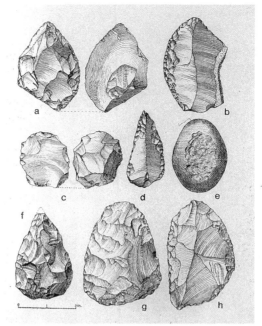

34   Middle Paleolithic (Mousterian) stone tools from Europe.

American.[20] And if that is the case, it surely suggests that Nean-derthals and moderns were one species and were in the process of evolving from one to the other in the Levant, they add.

This is not true, however. The species' highly similar tool kits can be explained, far more satisfactorily, as a reflection of their common ancestry, about 150,000 years previously. Both had evolved sepa-rately since that period, but in those days of sluggish cultural innova-tion, their tool technology had progressed slowly and in parallel. And that, very simply, is why they share the same level of implements. They had not yet evolved separately for long enough to develop dis-tinctive technologies. The problem of accounting for the stone tool kits' similarities is therefore more apparent than real. As Cambridge archeologist Paul Mellars puts it: "Early modern humans had no choice about being Middle Palaeolithic—the Upper Palaeolithic had not yet been invented!"[21]

In any case, judging a race or a tribe or a species purely from its tools is a notoriously tricky business. If you categorize Australian Aborigines this way, you might assume from their basic stone kits that they were like Neanderthals, for some of their implements are similar to those of the mid-Paleolithic period. Yet, while these people may use simple tools, they also have a store of some of the most sophisticated social and religious ideas that anthropologists have ever encountered, with their complex notions of stewardship of ritual land and their mythic creative period, The Dreaming, when spirits were believed to have shaped the land, bringing into being various species and establishing human life. You cannot see that complexity from pieces of stone.

And just because the same tools were shared by *Homo sapiens* and Neanderthals does not mean they were used in exactly the same way, a point stressed by Shea and his colleague Dan Lieberman, of Rutgers University.[22] These two researchers have tried to prize apart subtle distinctions between the two species' behavior from their skeletons and artifacts. Given that this fossil detritus was abandoned up to 100,000 years ago, their task is a daunting one, of course. As we have already pointed out, tens-of-millennia-old pieces of bone and stone tend to be mute on the subject of their owners' lifestyle. Nevertheless, the two scientists have produced some intriguing results.

For a start, they examined teeth of prey slain and left behind at both Neanderthal and modern human sites. (This quarry consisted mainly of gazelle, the Levant's Stone Age equivalent of the takeaway pizza.) Of particular concern for the two scientists were the alternating light and dark stripes exposed when a tooth is sliced open, banding that corresponds to layers put down in different seasons when distinctive foods were consumed. "This effect occurs because teeth cementum is produced by natural minerals interacting with collagen fibres," says Lieberman.[23] "When nutrition intake is poor, then collagen production drops and there is more mineral being packed into a tooth. It becomes darker as a result." In other words, when protein-poor winter food is eaten, dark bands occur, but with protein-rich summer foods, they are light.

To prove this point, Lieberman raised goats on separate low and high protein diets of hay and pellets, adding vitamins and fluorescent dyes at various stages. Then he removed their teeth, looked at the bands, and found, as expected, that when typical winter food is given, dark bands occur, but with summertime foods, they are light. This phenomenon therefore allows you to look at an ancient animal tooth, and tell in what season its owner died; dark outermost layers for a winter death, and light for summer. And when Lieberman examined gazelle teeth at Neanderthal sites in the Middle East he found there was a mixture of bandings, implying that the sites had been used at all times of the year. At modern human sites, banding was either light only, or dark only, indicating they had been used seasonally:

> Clearly modern humans migrated between various winter and summer locations, while Neanderthals lived in sites the year round. And that is crucial—for it's hard work to stay in the same place. The area becomes depleted of nuts, berries, tubers and vegetables, and the local game—even the gazelle—learn to avoid you. That is why hunter-gatherer tribes today move around all the time. Neanderthals, on the other hand, would have had to have travelled further and would have had to have worked harder to stay in the same place. And if you have to search for many hours for food, there would be no point in bringing back only a few berries or a couple of potatoes. You want a deer or a gazelle, a high protein shot in return for all your effort. It is the schlepp effect. You are not going to go schlepping about for long distances unless there is good reward. And in the Neanderthals' case that meant big game. They hunted more, but moved their homes less than modern humans. We moved more, but hunted less.

To establish this point, Shea and Lieberman looked for corroborating evidence, in this case in the quantities of the different implements which made up those ubiquitous Middle Paleolithic kits. And yes, both modern humans and Neanderthals used the same tools, but crucially, they had different favorites. Discarded stone pieces, the remnants of tool construction, divulged one key fact: Neanderthals were making, and jettisoning, hunting equipment at ten times the rate of modern humans. It was the same kit, but in different proportions, with spear points, often with impact damage indicating where they had struck bone, being found in profusion at Neanderthal sites. Lieberman concludes:

> The two species were living in the same environment but were exploiting it in different ways. Neanderthals were using it in a more resource intensive way compared with modern humans. They were working harder, and at earlier ages, than modern humans because they used the environment differently.

This approach would have had a profound impact on Neanderthals. Indeed, Lieberman believes the effect pervaded every bit of their lives, right down to their bones: "It is clear Neanderthals had thicker skeletons than modern humans, because their behaviour, not just their genes, was dissimilar. They were more active. They had to work harder and hunt more—because they used the environment differently from modern humans."

This idea led Lieberman to carry out experiments that—on anthropology's already lofty eccentricity scale—score commendably high ratings. In this case, he made armadillos run on treadmills for many hours a day. Now such a confluence, between jogging termite-eater and evolving human, may seem unlikely. The connection is real, nevertheless, for in his experiments, Lieberman found the more these animals exercise, the thicker their bones will be.

In fact, Lieberman has demonstrated this effect on several species. However, his favorite remains the plated, burrowing South American armadillo because females bear batches of four genetically identical offspring every pregnancy. Armed with two pairs of genetic doppel-gangers, Lieberman was therefore able to study the environment's effects on an armadillo's physiological development without worrying about the influence of inherited differences. He took two armadillos from a batch, and let them continue their normal snuffling lives. The other pair had to run on a treadmill for at least an hour every day. The results can now be seen in the sad-looking shoe-boxes of armadillo bones stored in Lieberman's cramped office. Armadillo joggers have noticeably thicker bones than genetically identical sedentary snufflers—observations that are borne out, scientifically, by Lieberman's careful autopsy measurements.

"Being big-boned is not merely a gift of the genes," he says. "If I want to make an animal hyper-robust, I can do it simply in the laboratory. I can make little Neanderthal armadillos—by making them run on treadmills when young." We can see this effect in humans today. For example, marathon runners not only develop powerful running muscles, but also thick bones to counter the stress of running so often.

And that, in turn, takes us back to the Neanderthals. Lieberman and Shea's work creates an evocative picture—of a species striving harder and harder to stay still. Lacking a less labor-intensive way to exploit their environment, they slogged towards hardship, big bones, athleticism, and extinction. But in *Homo sapiens*, we can see patterns of a lighter, more effective touch upon the environment.

There are other signs which divulge the distinctive nature of the Neanderthals' lifestyle. Lewis Binford, of Southern Methodist University, has studied several sites and has suggested that hunters— presumably male—brought no small mammal bones back to their family shelters: no rabbits, no foxes, and no rodents. Yet it is hard to imagine that they were not eating these creatures, argues Binford, since every other large carnivore does. The only likely explanation, therefore, is that they must have been eating them in the field. Only large pieces of cadaver—skulls or marrow bones—were returned to campsites, because they needed to be heated on the communal fire. In other words, Neanderthals killed and ate much more on their own. They were far less socially cohesive in their behavior than *Homo sapiens*.[24]

The net result of these various, coalescing influences was that by 40,000 years ago, the Neanderthals' heartland was being eroded by modern, socially cohesive humans spreading across Europe, just as the ice sheets fluctuated dramatically, constantly changing the climate and the environment of the region.

Ironically, it was under these conditions that "cold-adapted" European Neanderthals perished. They continued to live a relatively isolated lifestyle, as revealed by their flints which are hardly ever found much more than thirty miles from their source. By contrast, the incoming modern humans appear to have adopted wider and wider webs of social contacts—other "tribes" or groups—and established flourishing trades in flint, stone artifacts, and ornaments. By analyzing the sources of these implements, it is clear that raw materials were being transported up to two hundred miles, indicating a considerable commercial sophistication. In addition, structured campsites, storage pits, and primitive villages began to make their appearance. Neanderthals, for their part, continued to mark time, displaying little cultural evolution until they were almost extinct. They had little interest in exploration, at least compared to *Homo sapiens*, and never made boats, as far as can be determined, thereby missing the chance to colonize the islands of the Mediterranean, some of the choicest real estate to come on the market in Stone Age times, or easiest of all, cross the few miles of the Straits of Gibraltar to their ancestral homeland of Africa.

Homo sapiens, on the other hand, were forging widening alliances between relatives, trading partners, peers, and task groups, to judge from the shared artifacts and figurines which began to spread across the continent. These affiliations were to act as insurance policies in habitats where the only certainty was that resources would sometimes fail. Despite the stress of vacillating weather patterns, and a physique better suited to the tropics, modern humans showed an increasing ability to hold their place on this grim terrain. Neanderthals did not, and continued their ebb and flow of settlement as conditions changed. When circumstances got really tough, Neanderthals either died out locally or moved off, presumably in the expectation that they would return when the situation improved, in other words when the weather got better. But once *Homo sapiens* had established itself, Neanderthals would have found their old homelands had been taken over forever.

As a result, by about 35,000 years ago, Neanderthals had been reduced to living in isolated pockets in western Europe. Their access to good foraging and hunting ranges was reduced, their interchange of mates disrupted, and their ability to maintain population numbers seriously curtailed. The bleak hand of extinction was inexorably

35   Jean-Jacques Hublin in the cave of Zafarraya.

tightening its grip. By 30,000 years ago, they had all but vanished. Only a few regions on the high ground or the fringes of Europe (for example, caves in the Alps and in southern Spain) provided refuges for people that once flourished from Wales to Uzbekistan. Exactly where the last Neanderthal perished we will never be able to tell, though Zafarraya, near Málaga, remains as good a candidate as any.

Here, in a tiny, cramped passage in a limestone cavern etched into the parched Sierra de Alhama, twenty miles north of Torre del Mar, Spanish and French researchers have discovered remains which show Neanderthals were still lingering on less than 30,000 years ago, long after the species had apparently disappeared throughout the rest of Europe. "Southern Spain is the cul-de-sac of Europe, the very end of the continent," says one of the dig's directors, Dr. Jean-Jacques Hublin, of the Musée de l'Homme in Paris. "If Neanderthals were going to hang on anywhere, it would be here."[25]

For several years, Dr. Hublin—an Algerian-born, Paris-educated anthropologist who, like Yoel Rak, has displayed an enthusiasm for paleontological matters since childhood—has led teams of volunteers who have been meticulously scraping away thin layers of soil, using brushes and scalpels, within a narrow rock passageway that stretches for sixty feet into the limestone cliff above the village of Ventas de Zafarraya. Conditions are extraordinarily cramped, making other digs, such as Amud, look positively palatial. Only by working in shifts, and by assembling crude floors of planks under which individuals can excavate in three-foot-wide channels, has it been possible to explore the recesses of the cave. In this way, the team, directed by Hublin and his fellow overseer of excavation, Cecilio Barroso-Ruiz, has amassed a treasure trove of Stone Age detritus, flints, as well as human and animal remains. Cooked goat meat seems to have been a particular culinary favorite among Neanderthals here, for the researchers found piles of bones of the local subspecies, *Capra ibex pyrenaica*. They also found the remains of at least one hearth.

Before Zafarraya was excavated, the last Neanderthals were known from Saint-Césaire and Arcy-sur-Cure in France where 36,000-year-old and 32,000-year-old remains have been discovered respectively. However, using radiocarbon and uranium-thorium measuring techniques at Zafarraya, scientists have now pushed forward that date by more than 2,000 years.[26] "Quite clearly Zafarraya was used by Neanderthals for a long time after we thought they had been rendered extinct," adds Dr. Hublin.

But what has caused the real surprise is the nature of the stone implements left behind in the cave. They are Middle Paleolithic. Elsewhere in Europe these basic scrapers and knives had been replaced by Aurignacian tools—named after the site of their discovery, Aurignac in southern France. This kit was far more sophisticated in nature and first appeared about 40,000 years ago. It is uniquely associated with *Homo sapiens*, and is characterized by its long retouched blades; short, steep-sided scrapers; and bone points. Until this time, bone tools had rarely been made. With the arrival of the Aurignacian kit they became common in Europe.

The Aurignacian kit reveals an entirely new way of working stone, and demonstrated a deeper, more complex form of thinking. A Neanderthal making a Middle Paleolithic stone tool would simply pick up a lump of flint, and strike it with another stone until a hand axe or spear point had been shaped. But when a modern human craftsman began his Aurignacian handiwork, he or she would strike down at the top of the flint block, shaving off many flint slivers which would ultimately have many purposes—scrapers, knives, spear points, engraving tools, piercers, and much more—betraying the presence of a far more complex mental template, one that clearly envisaged many simultaneous options in a single act. Neanderthals basically exhibited only one. They created a simple Paleolithic penknife. Modern humans produced a Stone Age Swiss Army knife.

"Once modern humans arrived, all sorts of different, more advanced, tools were made," adds Dr. Hublin, "even by Neanderthals, and that has confused our understanding of their behavior and of the reasons for *Homo sapiens'* success." For example, at a Neanderthal site at Châtelperron in mid-France, archeologists uncovered tools that are far more sophisticated in composition and construction than the normal Middle Paleolithic fare, and have some similarities with Aurignacian implements. "However, it now seems clear that Neanderthals only made advanced tools in areas such as Châtelperron that were close to modern human habitation," says Hublin:

> In other words, they got their ideas from our ancestors. Neanderthals were hunters, and I guess when they saw modern humans using bone-pointed spears to kill animals they would quickly have understood how to make the same objects. Similarly, it was not until modern humans appeared in the Middle East that Neanderthals began to bury their dead.

In short, Neanderthals may only have got this culture from modern humans, for in southern Spain, where *Homo sapiens* were conspicuous by their absence until after 30,000 years before present, they continued to make only Middle Paleolithic tools. In other words, in the rocky lair of Zafarraya, the Neanderthals survived in this biological and cultural refuge until those African parvenus, *Homo sapiens*, finally spread down from France and northern Spain, bringing their high-tech stone wares. The result was extinction for the last of the first Europeans.

An evocative glimpse of the last days of Neanderthal life is provided by zoologist Jonathan Kingdon in his book, *Self-Made Man*:

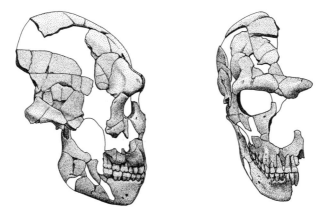

36 The Saint-Césaire Neanderthal from France, dated at about 36,000 years, is one of the most recent known.

Neanderthals . . . survived the winter nights by lighting fires in caves which were close to concentrations of animals. A meat harvest sufficient to see a few individuals throughout the winter was best ensured by small family units living in, and presumably defending, their own caves or shelters. These were sited on animal routes or gathering grounds. A very few individuals killing, butchering and moving large carcasses had to expend more energy and possess greater strength and endurance than members of a large group. The Neanderthals saw through the winter shortages by being few and far between, living within their means and taking a mixed bag of the available animals: cave bears, mammoths, rhinos, horses, deer, reindeer, bison, and ibex. They relied upon being strong, resilient and cooperative within the family, but there is no evidence that they gathered in any numbers.[27]

Neanderthal populations would have been close to starvation during many severe winters, suggests Kingdon. Without other hominids about, these vicissitudes could be absorbed but with the appearance of modern humans, assisted by their greater social flexibility and organization, their retreat and extinction followed all too rapidly. Cro-Magnons would have monopolized the goats, deer, and other game that was then available, and Neanderthals would have been left without meat, the mainstay of their diets in those harsh times. Starvation would have been inevitable. It was our arrival that brought about their end, directly or indirectly.

It is a poignant tale, and no matter how carefully it is related, it is hard not to judge it in terms of evolutionary "promotion." Because Neanderthals could not adjust to a changing world as well and as quickly as did *Homo sapiens*, it is popularly imagined that they must have been inferior and therefore must have been replaced by something "better"—i.e., modern human beings. That is an interpretation of evolution as a single file march of progress, "the one picture immediately grasped and viscerally understood by all," as Stephen Jay Gould puts it. In *Wonderful Life*, Gould stresses the misleading nature of this vision. "Life is a copiously branching bush, continually pruned by the grim reaper of extinction, not a ladder of predictable progress." [28]

The trouble is that, in seeking the causes of Neanderthal extinction, a story that we follow with fascination because it helps us understand the subtle, but critical differences between them and us, we are by definition emphasizing their failures. We should also remember the species flourished for a quarter of a million years and demonstrated extraordinary sophistication: they buried their dead; families took care of weaker members (as far as we can see from the bones of their elderly and disabled, such as those found at Shanidar); there are even traces of ornaments at the last Neanderthal sites—pendants with holes for string—which they either made themselves, or obtained through trade with modern humans. They had brains as large as, and occasionally larger than, those of *Homo sapiens*, and there is no evidence that individuals were necessarily less gifted than their modern human counterparts. They had enough sense, skill, and organization to survive 200,000 bitter European winters, and held on in the continent long after most of the rest of the world succumbed to the passage of *Homo sapiens*. Modern humans had reached Australia before they eventually made their way into the Neanderthal heartland of modern France and Germany, a process that took about 10,000 years. By contrast, it took *Homo sapiens* only a few millennia

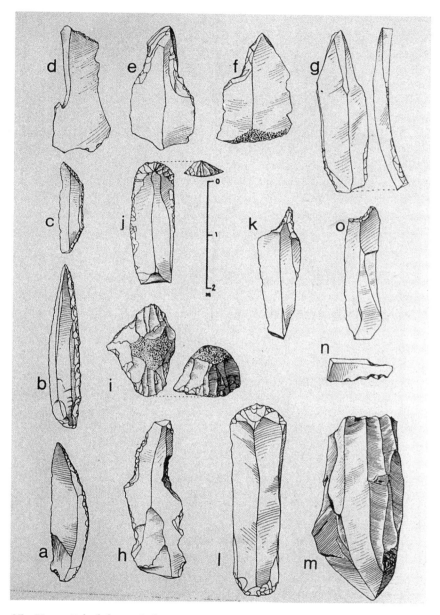

37 Upper Paleolithic tools from Europe. All were the product of modern humans except for (a), which was made by a French Neanderthal.

to conquer the Americas, from Alaska (then connected to Russia) in the Arctic Circle, to the most southern tip of South America, at Tierra del Fuego, near the Antarctic Circle, though these two continents were, admittedly, unpopulated. Neanderthal-occupied Europe was clearly a much harder nut.

In the end, the species probably perished because its social groups were small, as Kingdon has pointed out, so that when times were hard, during harsh climatic changes when food was scarce, *Homo sapiens* were simply able to call on better organizational backup, and share ideas with a bigger group of individuals. We had the big battalions.

We should not be overcome with a sense of our own organizational superiority, however, for it is quite wrong to equate survivorship with some form of worth. Extinction is the inevitable destiny of all evolutionary lineages. Virtually every fossil that we now know about has no modern descendant. They flowered on the bush of life, and then withered without offspring. Consider the case of the early primates, the creatures from which we are descended. It is estimated from the fossil record that there must have been 6,000 different species, but only 185 are alive today. Most were evolutionary dead ends, because the demise of a species is the rule, not the exception, in biology, and we should not think that human beings are exempt from its harsh inevitability. "For every species alive today, a hundred now lie frozen into the rocky sediments of the earth," as Erich Harth, of Syracuse University, New York, puts it with some understatement.[29] In other words, time simply ran out for the Neanderthals—just as it will for *Homo sapiens* one day. We should therefore recall the words of Ecclesiastes: "The race is not to the swift, not the battle to the strong, neither ye bread to the wise, not yet riches to men of understanding, not yet favor to men of skill; but time and chance happeneth to them all."

Opportunity knocked for *Homo sapiens* only a relatively few millennia ago—that is our good fortune, at present. In the next chapter, we shall examine the final proof that has emerged—not from the bones of the dead, but from the blood of the living—to show the singular nature of this act of evolutionary providence.

5

The Mother of All Humans?

Fossil bones and footsteps and ruined homes are the solid facts of history, but the surest hints, the most enduring signs, lie in those minuscule genes. For a moment we protect them with our lives, then like relay runners with a baton, we pass them on to be carried by our descendants. There is a poetry in genetics which is more difficult to discern in broken bones, and genes are the only unbroken living thread that weaves back and forth through all those boneyards.

Jonathan Kingdon, Self-Made Man and His Undoing

Picture this simple scene. A group of lowland gorillas—a male, six females, and their offspring—amble through a forest in central Africa. Moving on all fours, their knuckles bearing the weight of their powerful upper bodies, the animals pluck at the occasional shoot, rummage for berries, and gnaw at bits of vegetation. Then they stumble into a clearing, where they are confronted by another gorilla group, similarly led by a dominant silverback. The two males stare at each other. Then they start to display. They roar, throw leaves in the air, beat their chests, and finally begin running sideways, tearing up bushes and plants and banging their fists on the ground. The demonstration proves too much for the leader of the first group. He turns, and, with his females and children in tow, scrambles back into the forest.

It is a typical, though not particularly frequent, gorilla encounter. Despite their fearsome reputation, these animals—the largest of all the primates, and sharers, with chimpanzees, orangutans, gibbons,

and humans, of the hominoid (as opposed to hominid) ape classification—are peaceful, browsing vegetarians. They hardly ever indulge
in serious violence and have only occasional skirmishes. Even the
best of families can fall out, of course.

However, there is one striking feature about these two groups (and
about all other gorillas in our forest) that sheds a great deal of light,
not upon our primate cousins, but upon ourselves. If you took specimens of a special type of genetic material, known as mitochondrial
DNA, from those two squabbling males, and then compared them
with samples from an Eskimo and an Australian Aborigine, you would
uncover a surprising fact. You would find the latter pair (the humans)
are more genetically alike than the former (the gorillas). Yet the
Eskimo and Aborigine live half a world away from each other in
dramatically contrasting environments. Our two gorilla combatants
share the same forest. Nevertheless, there is more variation in their
genetic constitutions than the most distantly related members of
Homo sapiens. When it comes to genes, "humans are less different . . .
even than lowland gorillas living in a restricted geographic area of
west Africa," state a team of Harvard University anthropologists, led
by Professor Maryellen Ruvolo, in a paper in the *Proceedings of the
National Academy of Sciences*[1] in 1994. Nor is this phenomenon
restricted to *Homo sapiens* and *Gorilla gorilla gorilla* (as the lowland
gorilla is imaginatively classified). The Harvard team's research on
chimpanzee and orangutan mitochondrial DNA has also revealed
these species to be considerably more diverse than *Homo sapiens*.

It is not the gorilla, nor the chimpanzee, nor the orangutan, that is
unusual, however. Each enjoys a normal spectrum of biological variability. It is the human race that is odd. We display remarkable geographical diversity, and yet astonishing genetic unity. This dichotomy
is perhaps one of the greatest ironies of our evolution. Our nearest
primate relations may be much more differentiated with regard to
their genes but today are consigned to living in a band of land across
Central Africa, and to the islands of Borneo and Sumatra. We, who
are stunningly similar, have conquered the world.

This revelation has provided the unraveling of our African origins
with one of its most controversial chapters. And it is not hard to see

why. The realization that humans are biologically highly homogeneous has one straightforward implication: that mankind has only recently evolved from one tight little group of ancestors. We simply have not had time to evolve significantly different patterns of genes. Human beings may look dissimilar, but beneath the separate hues of our skins, our various types of hair, and our disparate physiques, our basic biological constitutions are fairly unvarying. We are all members of a very young species, and our genes betray this secret.

It is not this relative genetic conformity per se that has caused the fuss but the results of subsequent calculations which have shown that the common ancestor who gave rise to our tight mitochondrial DNA lineage must have lived about 200,000 years ago. This date, of course, perfectly accords with the idea of a separate recent evolution of *Homo sapiens* shortly before it began its African exodus about 100,000 years ago. In other words, one small group of *Homo sapiens* living 200 millennia ago must have been the source of all our present, only slightly mutated mitochondrial DNA samples—and must therefore be the fount of all humanity. Equally, the studies refute the notion that modern humans have spent the last one million years quietly evolving in different parts of the globe until reaching their present status. Our DNA is too uniform for that to be a realistic concept. As Professor Ruvolo's team points out, their research places a common human ancestor "at 222,000 years, significantly different from one million years (the presumed time of an *Homo erectus* emergence from Africa). The data . . . therefore do not support the multiregional hypothesis for the emergence of modern humans." In fact, Professor Ruvolo puts it more emphatically when not constrained by the normal, dry academic prose required of a scientific journal. "No way does any of this information support the idea [that] there was a common human ancestor living as long ago as one million years," she says. "In fact, when I wrote that paper in late 1993, I assumed a date of one million years before present for the emergence of *Homo erectus* from Africa. But now that date has been pushed back to at least 1.8 million years ago following those new dates of the *Homo erectus* skeletons in Java. Our research therefore rejects the multiregional hypothesis even more strongly." On the other hand, the

Harvard team's work is entirely consistent with the Out of Africa theory. "It is a comparative way of showing you how humans are all very, very alike—and that similarity translates into one thing: the recency of the origin of our common ancestor," adds Professor Ruvolo.

Not surprisingly, such intercessions into the hardened world of the fossil hunter, by scientists trained in the "delicate" arts of molecular biology and genetic manipulation, have not gone down well in certain paleontological circles. The old order has reacted with considerable anger to the interference of these "scientific interlopers." The idea that the living can teach us anything about the past is a reversal of their cherished view that we can best learn about ourselves from studying our prehistory. Many had spent years using fossils to establish their interpretations of human origins, and took an intense dislike to being "elbowed aside by newcomers armed with blood samples and computers," as *The Times* (London) put it. "The fossil record is the real evidence for human evolution," announced Alan Thorne and Milford Wolpoff in one riposte (in *Scientific American*)[2] to the use of mitochondrial DNA to study our origins. "Unlike the genetic data, fossils can be matched to the predictions of theories about the past without relying on a long list of assumptions." Such a clash of forces has, predictably, generated a good many sensational headlines, and triggered some of the most misleading statements that have ever been made about our origins. Scientists have denied that these genetic analyses reveal the fledgling status of the human race. Others have even rejected the possibility of ever re-creating our past by studying our present in this way. Both views are incorrect, as we shall see. Even worse, the multiregionalists have attempted to distort the public's understanding of the Out of Africa theory by deliberately confusing its propositions with the most extreme and controversial of the geneticists' arguments. By tarnishing the latter they hope to diminish the former. This chapter will counter such propaganda and highlight the wide-ranging support for our African Exodus provided, not just by the molecular biologists, but by others, including those who study the words we speak and who can detect signs of our recent African ancestry there. We shall show not only that the majority of leading evolutionists and biologists believe in such an idea but that their views raise such serious questions about the multiregional hypothesis that its future viability must now be very much in doubt.

Unraveling the history of human migration from our current genetic condition is not an easy business, of course. It is a bit like trying to compile a family tree with only an untitled photograph album to help you. "Our genetic portrait of humankind is necessarily based on recent samplings, [and] it is unavoidably static," says Christopher Wills of the University of San Diego. "Historical records of human migrations cover only a tiny fraction of the history of our species, and we know surprisingly little about how long most aboriginal people have occupied their present homes. We are pretty close to the position of a viewer who tries to infer the entire plot of *Queen Christina* from the final few frames showing Garbo's rapt face."[3]

It is an intriguing image. Nevertheless, biologists are beginning to make a telling impact in unraveling this biological plot and in understanding *Homo sapiens'* African exodus. And they have done this thanks to the development of some extraordinarily powerful techniques for splitting up genes, which are made of strands of DNA (deoxyribonucleic acid) and which control the process of biological inheritance. By dividing DNA's filaments of individual chemical bases—adenine (A), cytosine (C), guanine (G), and thymine (T)—into small sections, and by making millionfold copies of these pieces, they can study DNA in a way that easily differentiates one person from another, permitting great strides to be made in medical research—for instance, in pinpointing the causes of inherited diseases. However, this technology can also be used to exploit DNA, not for its medical knowledge, but as a bearer of information, a courier from the past, and a very informative one at that.

"The most golden of molecules,"[4] DNA is found in two different places within our bodies. There is mitochondrial DNA, and there is nuclear DNA. The latter makes up the genes that control the development of the growing embryo, and determine whether we will be short or tall, blue- or brown-eyed, and much else. This type of DNA is found in the nucleus of every cell in our body (hence the "nuclear" adjective) and is bundled together into chromosomes, along which are ranged those genes for brown eyes, height, and other attributes.

In total, we have twenty-three pairs of chromosomes numbered from one to twenty-two, plus either a couple of X chromosomes, or a Y and an X. The former combination dictates that a person will be a woman while the presence of a Y chromosome ordains masculinity.

When a cell divides, so does its DNA. Its double helical strands of complementary chains of A, C, G, and T bases separate, and each grows a new second chain, with the result that an exact copy of the originator's genetic script is created. This duplicate DNA—in the form of a new set of twenty-three pairs of chromosomes—migrates to the freshly created cell where it renews the business of directing the manufacture of proteins, the biological building blocks from which our bodies are constructed. Just six million millionths of a gram of DNA carries as much information as ten volumes of the complete *Oxford English Dictionary*. Thanks to this marvelous biological lexicon, a single cell, the fertilized egg, created when egg meets sperm, that joins the inheritance of a mother and a father in equal parts, is gifted with the power to form a unique individual. The human body is made up of a hundred million million cells of many different sorts—those of blood, skin, bone, kidneys, lungs, the brain, and many others—and all contain the inherited information, carried in DNA, that comes from that first, single cell.[5] We shall examine the importance of nuclear DNA research in the unraveling of our history later in this chapter. Before we do, however, let us look at the far more controversial role played by the other form—mitochondrial DNA.

This second type of genetic material is found outside the nucleus, but inside the cell, in objects called mitochondrial organelles, which act as cells' microscopic power packs, and which have their own genetic blueprint: mitochondrial DNA. However, mitochondrial DNA differs from nuclear DNA in one important aspect. Unlike the latter, bequeathed as a fifty-fifty mixture from both our parents, the former is inherited solely from our mothers, because an unfertilized egg needs to be crammed with energy-providing mitochondrial organelles to sustain the embryo. Sperm only carries mitochondrial DNA in its tail, which is left behind when fusion takes place with the egg. As a result, a male makes no mitochondrial contribution to its offspring, and a person always inherits his or her mother's mitochondrial DNA. In a man, this inheritance is spliced out of the business of reproduction. By contrast, a woman inherits her mother's mitochondrial DNA, and her mother's mother's, and her mother's mother's mother's, and so on back into the mists of humanity's prehistory. Think of mitochondrial DNA as being a little bit like Jewishness. Both are inherited through the maternal line.

Having an unbroken biological bond with our past is clearly a source of important information. However, we should not assume we have exactly the same mitochondrial DNA as our great-grandmother twenty times removed. The thread, although not broken, is frayed very slightly over the millennia because it accumulates occasional mutations. Imagine a DNA sequence made up of a long stretch of those four A, C, G, and T bases that we mentioned earlier. Sometimes a mistake in replication occurs, and a C replaces a G, or an A is substituted for a T. It is a well-understood phenomenon, which, in the nucleus, is usually spotted and put right by special biological repair molecules. In the organelles, the mechanism for reconditioning old DNA is much less effective, so mutations accrue at a more rapid rate.

This apparent evolutionary oversight is good news for the biologist. Thanks to the development of those highly specific techniques for studying DNA that we mentioned earlier, they can examine bases along a particular stretch of mitochondrial DNA in different people, from different races, and count those bases that are shared, and those that are not. And the greater the number of unshared bases, the greater has been the number of accumulated mutations, and the longer it has been since the two individuals (and presumably the populations they represent) shared a common ancestor. The fewer the number of mitochondrial DNA differences, the greater the similarity, and the more recent must have been the date that they shared a forebear. In this way, scientists realized they could study the relatedness of the world's dispersed peoples.

And this is exactly what Allan Wilson, Rebecca Cann, and Mark Stoneking, working at the University of California, Berkeley, did in 1987.[6] They took specimens from placentas of 147 women from various ethnic groups and analyzed each's mitochondrial DNA. By comparing these in order of affinity, they assembled a giant tree, a vast family network, a sort of chronological chart for mankind, which linked up all the various samples, and therefore the world's races, in a grand, global genealogy.

The study produced three conclusions. First, it revealed that very few mutational differences exist between the mitochondrial DNA of human beings, be they Vietnamese, New Guineans, Scandinavians, or Tongans. Second, when the researchers put their data in a computer and asked it to produce the most likely set of linkages between the different people, graded according to the similarity of their mitochondrial DNA, it created a tree with two main branches. One consisted solely of Africans. The other contained the remaining people of African origin, and everyone else in the world. The limb that connected these two main branches must therefore have been rooted in Africa, the researchers concluded. Lastly, the study showed that African people had slightly more mitochondrial DNA mutations compared to non-Africans, implying their roots are a little older. In total, these results seemed to provide overwhelming support for the idea that mankind arose in Africa, and, according to the researchers' data, very recently. Their arithmetic placed the common ancestor as living between 142,500 and 285,000 years ago, probably about 200,000 years ago. These figures show that the appearance of "modern forms of *Homo sapiens* occurred first in Africa" around this time and "that all present day humans are descendants of that African population," stated Wilson and his team.

The Berkeley paper outlining these findings was published in the journal *Nature* in January 1987, and made headlines round the world, which is not surprising given that Wilson pushed the study's implications right to the limit. He argued that his mitochondrial tree could be traced back, not just to a small group of *Homo sapiens*, but to one woman, a single mother who gave birth to the entire human race. The notion of an alluring fertile female strolling across the grasslands of Africa nourishing our forebears was too much for newspapers and television. She was dubbed "African Eve"—though this one was found, not in scripture, but in DNA. (The honor of so naming this genetic mother figure is generally accorded to Charles Petit, the dis-

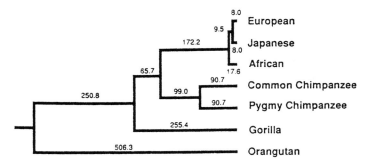

38 Horai's mtDNA tree is based on complete sequences from both apes and humans (see pages 131–32). Note the shallow separation of the three human samples.

tinguished science writer of the *San Francisco Chronicle.* Wilson claimed he disliked the title, preferring instead, "Mother of us all" or "One lucky mother.")[7]

The image of this mitochondrial matriarch may seem eccentric but it at least raises the question of how small a number of *Homo sapiens* might have existed 200,000 years ago. In fact, there must have been thousands of women alive at that time. The planet's six billion inhabitants today are descendants of many of these individuals (and their male partners), not just one single super-mother. As we have said, we humans get our main physical and mental characteristics from our nuclear genes, which are a mosaic of contributions from myriad ancestors. We appear to get our mitochondrial genes from only one woman, but that does not mean she is the only mother of all humans.

"Think of it as the female equivalent of passing on family surnames," states Sir Walter Bodmer, the British geneticist.[8] "When women marry they usually lose their surname, and assume their husband's. Now if a man has two children, there is a 25 percent chance both will be daughters. When they marry, they too will change their name, and his surname will disappear. After twenty generations, 90 percent of surnames will vanish this way, and within 10,000 generations—which would take us from the time of 'African Eve' to the present day—there would only be one left." An observer might assume that this vast, single-named clan bore a disproportionately high level of its originator's genes. In fact, it would contain a fairly complete blend of all human genes. And the same effect is true for mitochondrial DNA (except of course it is the man who is "cut out"). The people of the world therefore seem to have basically only one mitochondrial "name." Nevertheless, they carry a mix of all the human genes that must have emanated from that original founding group of *Homo sapiens*. It is a point that Wilson tried, belatedly, to make himself. "She wasn't the literal mother of us all, just the female from whom all our mitochondrial DNA derives."

But there were other criticisms in store for the Berkeley team. For a start, of the 147 individuals who had been used to supply the raw data, 98 had been found in American hospitals. And in particular, of the 20 "Africans," only two were actually born there. The other 18 were African-Americans, though they were classified as Africans for the study. Given that so much had been made of its African results, the failure to sample directly from people who actually lived on the continent appeared a little remiss. For their part, the researchers pointed out that technological constraints forced this local arrangement upon them. "The techniques that we had to use then required a great deal of mitochondrial DNA and we could only get that from placentas," recalls Mark Stoneking.[9] "There was not enough in a normal blood sample. Indeed, we would have bled our subjects dry fulfilling our needs." So the team had to restrict themselves to using placentas from the San Francisco Bay area. In any case, added the researchers, this geographical inexactitude made little difference. Until very recently, male African-Americans did not often produce children with white women. Interracial breeding was almost exclusively between black women and white men. The African origins of the resulting offspring's mitochondrial DNA (from their mothers) would therefore have been preserved.

And finally there was the issue of the regularity of mutation accretion. The Berkeley team had found there was an average 0.57 percent divergence between the various DNA samples in their survey. In other words, they found that along a stretch of 1,000 bases of mitochondrial DNA, the average number of A, C, G, and T changes between two different people from the whole sample was 5.7. They then assumed that two samples would differ by twenty to forty base mutations every million years (we shall see why in a couple of pages) giving a divergence mutation rate of 2 to 4 percent per million years. But which is the correct rate? And are these mutations acquired at a roughly steady rate—say 2 percent per million years? Or are they gathered more rapidly at some periods compared with others? If the latter case prevails, great care should be taken when reading this molecular clock, some scientists cautioned, a warning that made little impact on Allan Wilson, who could claim, quite reasonably, to be one of the world's greatest experts on DNA mutation rates. Working with Vincent Sarich in the 1960s, he had carried out pioneering research that had questioned one of the most venerable precepts of paleoanthropology: that apes and humans had diverged as separated lines an extremely long time ago, probably between fifteen to thirty million years ago.

Fossils, the only "real evidence for human evolution," revealed that this ancient division had to be true, claimed paleontologists. Sarich and Wilson showed otherwise, by comparing the structure of proteins in apes and humans.[10] The manufacture of proteins is directed by DNA and mutations in the latter will therefore be revealed by slight changes in the former, the two reasoned. So they looked at proteins called serum albumins in both apes and humans and found their structures were remarkably similar—far too close to justify an ancient twenty-million-year-old division. It was more like five million, said Sarich and Wilson. Subsequent tests, using a battery of techniques, including protein electrophoresis, amino acid sequencing, restriction mapping of mitochondrial DNA, and sequencing of mitochondrial and genomic (nuclear) DNA, have produced virtually the same answer for the chimp-human division. "We were variously ignored, abused and scorned," recalls Sarich.[11] "But look at the figure they talk about now. We were more or less right. If you need better evidence that there is a molecular clock working in there somewhere, I don't know what it is." Score one to the geneticists.

In fact, paleontologists had made a mistake in classifying fragmentary jawbones and teeth which had suggested to them that the ancient fourteen-million-year-old bones of the apes called *Ramapithecus*, found in India and Pakistan, and another named *Kenyapithecus* from East Africa, were actually the ancestors of the hominid line that gave rise to *Homo sapiens*. But as we saw in Chapter 2, these were actually related to other ape lines. The fossil hunters were forced to recant, making a mockery of Thorne and Wolpoff's denigration of genetic data compared to the fossil record as the only "real evidence" for human evolution. Evaluating bones and pieces of skull also relies on assumptions that may be overturned.

So, armed with these new genetic findings, scientists have been able to calculate how much genetic variation there is between chimps and humans, and so determine the mutation rate for mitochondrial DNA in primates. This gives us our benchmark for working out when *Homo sapiens'* common ancestor lived, and provides the divergence mutation rate of 2 to 4 percent, as well as the range of dates of 142,500 to 285,000 years before present for the appearance of Eve.

However, the other complaints about Wilson's survey could not be ignored so easily. So the Berkeley team repeated their research, introducing several changes in methodology, and in 1991 published two key papers (in *Proceedings of the National Academy of Sciences*, and secondly in *Science*)[12] shortly before Wilson's death from leukemia. These demonstrated how the researchers had employed more detailed analysis—by sequencing, in full, two segments of the control region, called the hypervariable subregions, a highly mutable part of the whole mitochondrial DNA complement. (In their first survey, readers will recall, they broke up mitochondrial DNA into rather crude blocks of bases.) It was also based on a more reliable ethnic mix as a source of samples. Again their work produced those two branches which place mankind's birthplace firmly, and recently, in Africa. "Our study provides the strongest support yet for the placement of our common mitochondrial DNA ancestor in Africa some 200,000 years ago," they announced.

The issue seemed quite settled. With an absence of ancient mitochondrial DNA in people today, the study showed our ancestors must have emerged out of Africa and completely supplanted, without any interbreeding, existing populations of other human lines, an extreme form of replacement that was not an essential of the original African Exodus model put forward by scientists such as Chris Stringer and Günter Bräuer. This allowed the possibility of limited interbreeding. There was not a single sign of that in the genetic data, however.

Everything seemed nicely resolved—until scientists began to look more closely at the statistics used by Wilson, Cann, and Stoneking. Then one or two of them became uneasy. For example, Maryellen Ruvolo noticed that there were actually two versions of the African tree created in the Berkeley's later work. In the 1991 PNAS paper, !Kung bushmen from southern Africa were found at the base of the two main branches, suggesting their ancestors could actually be the progenitors of all humanity. In the second 1991 paper, Pygmies were promoted to this hallowed position. "You couldn't have it both ways on the same set of data," recalls Ruvolo. So she prepared a paper for *Science* pointing out the shortcomings of the Berkeley team's work, only to find she had been beaten to publication by Alan Templeton, a geneticist at Washington University, St. Louis, and—ironically—Mark Stoneking.[13] The latter, alerted to the statistical blemishes in his own group's study, admitted its flaws. He did not feel they entirely invalidated the Berkeley work, however. "The data from mitochondrial DNA is consistent with, but does not prove, that modern humans had a recent African origin," he said.

By contrast, Templeton was far more outspoken in his criticisms, which he outlined in brief in *Science* in 1992, and then in full in *American Anthropologist* in 1993.[14] He denounced the very concept of assuming that a gene tree was the same as a population tree. The former reflects the evolutionary history of a particular piece of DNA, he said; the latter indicates the movements of entire groups of individuals, and all the genes these groups carry. "My recent ancestors came from Scotland, Ireland, Germany, and the Netherlands," explains Templeton:

> My recent genetic roots are spread out over several countries and are not limited to one geographical location. Yet any particular gene I have will trace back to only one of these countries. For example, my mitochondrial DNA definitely came from Germany, the origin of my maternal grandmother's mother. However, my Y chromosome—inherited from father to son—

came from Scotland where my paternal grandfather lived. Hence, different genes can have different geographical origins. This is why gene trees and population trees are not necessarily the same thing. The Wilson group automatically assumed their tree was a population tree, when it was in fact a gene tree.

In other words, you could sum up Templeton's case quite neatly in the words of the Woody Guthrie song, "I'm from everywhere, man."[15]

But far more damning than this qualitative critique was Templeton's quantitative attack on the Berkeley team's computing methods. The researchers used a program called Phylogenetic Analysis Using Parsimony (PAUP). They put in their data, and out popped the tree. "As a result of analysing just one run, they fooled themselves into thinking they had a well-resolved evolutionary tree," adds Templeton. "In fact, there are thousands of equally good but different trees that can be made from the Berkeley data."

This point is acknowledged by Ruvolo, who accepts the research is flawed, but unlike Templeton does not believe that "Eve is officially dead." She admits thousands of different trees can be grown from the data, but points out that nearly all these mitochondrial bushes are only trivially different from each other. "In fact, we found three groups of trees—although there may be more if one searched further. Two had their roots in Africa, while the third's origins are unclear. So there is still evidence of an African origin, but it is not proof."[16]

Such qualifications have largely gone unheard, however. Such was the robustness of the original hype about Eve, and so emphatic the subsequent criticism of her existence, that most observers could reasonably assume that her rise was temporary and her fall permanent. She had been cast out of her African Eden forever. And she was not alone. Eve was popularly associated, in many people's minds, with the Out of Africa theory—a connection the multiregionalists specifically, and misleadingly, tried to reinforce. (In their *Scientific American* paper, Thorne and Wolpoff continually referred to the ideas of Stringer, Bräuer, and the others as the Eve hypothesis, despite the fact that the latter group had developed their Out of Africa model on fossil evidence and merely viewed the genetic study by the Berkeley group as a provider of welcome support.)[17] The fall of one therefore seemed to imply the fall of both. In fact, the latter retains a ruddy sheen of intellectual good health, for it does not depend on the work of Wilson, Cann, and Stoneking. In any case, Eve had by no means been expelled from her mitochondrial paradise. "To paraphrase Mark Twain," says Roger Lewin, in *The Origin of Modern Humans*, "reports of Eve's death have been greatly exaggerated." There was much in the two Berkeley studies that still suggested, but did not prove, that we had only emerged from our African homeland a short time ago on our road to world domination. For one thing, establishing the fact that most Africans are mitochondrially more disparate than the rest of the world's population is significant, for all the gleeful trumpeting by Milford Wolpoff and others. Nor have the Berkeley team been the only ones to demonstrate such diversity. A large-scale mitochondrial analysis of more than 3,000 people, carried out by Andrew Merriwether of the University of Pittsburgh, Douglas Wallace of Emory University, Atlanta, and a group of other geneticists, revealed, in 1991, that "the native African populations have the greatest diversity and, consistent with evidence from a variety of sources, suggests an African origin for our species."[18]

More to the point, the question of mapping the roots of our African Exodus is not actually the most important when considering modern mankind's origins. We know we come from Africa. The dispute is about whether we did so very recently, within the last 100,000 years, and about whether we replaced all other forms of humans in the process, or whether different races today have far more ancient antecedents. "It is not really a question of where our ancestor arose," Stoneking acknowledges. "Even if we are able to prove beyond any statistical doubt that there was an African origin, it would not distinguish between the two competing hypotheses. The question is: When did we arise?"[19]

And this is a problem. As we have seen, scientists calibrate their mitochondrial DNA clocks by counting the number of mutations that divide the various races and compare these figures with those found in the chimpanzee from whom we diverged about five million years ago. As Jonathan Kingdon so neatly puts it: "The great African apes are genetic Rosetta Stones whereby we get the measure of our humanity."[20] Unfortunately, there is still some disagreement about when humans and chimpanzees started evolving separately, despite Wilson and Sarich's work. Some say as little as four million years ago, and some as much as eight million, though no one seriously adheres to the old twenty-million-year figure. Which date you accept greatly affects the outcome of your final calculation. Many scientists believe four to six million is about right, however.

But there are other headaches that beset the mutation counter, such as the issue of multiple substitutions. As we have seen, an A base can replace a G base during DNA replication, and that counts as a single mutation. But several generations later, that A might be replaced by a C. To the scientist, it would only appear as if one mutation separated the two samples. In fact, two would have occurred. Or an A might mutate to a C, and then back to an A—in which case, no mutations would appear to have occurred. "There are mathematical formulas that take account of this, but they are not perfect," admits Stoneking. "They make statistical assumptions that can be questioned, though obviously I believe they are good enough." Nevertheless, the making of any supposition leaves the geneticists open to the accusation that they are slightly doctoring their equations to suit the desired outcome.

One solution to this problem has been adopted by Maryellen Ruvolo. Her work on gorilla, chimp, and human DNA, which we encountered at the beginning of the chapter, exploits a less mutation-prone piece of mitochondrial DNA, known as the COII gene. Because there are fewer mutations piling up on this section, the problem of multiple substitutions is not a serious issue. And, as we saw, her research also produces a figure of around 200,000 years for the age of the common ancestor. In showing that gorillas and chimpanzees are very highly genetically differentiated she is, as she puts it, demonstrating that "long, ancient mitochondrial lineages are a dime a dozen in the primate world. In gorillas, in chimpanzees and in orangutans, they are perfectly routine for example. It just so happens that *Homo sapiens* does not have them. So something weird must have happened in humans. We are the glaring exceptions—because we are a very recently expanded species."[21]

There is an additional critical factor that is often ignored, however—researchers may one day stumble upon a remote, as yet unrealized group of people whose mitochondrial DNA sequences are wildly divergent from the rest of humanity's. Such a race would utterly distort the comfortable genetic statistics that currently support the Out of Africa theory, for its existence would show *Homo sapiens* had been evolving for far longer, and had acquired more mutations, than science had previously suspected. Our lineage would therefore be forced back much further, possibly even into a domain that might resurrect the multiregion hypothesis. It is a possibility, though it is an unlikely one as Masami Hasegawa and Satoshi Horai, of Japan's National Institute of Genetics, point out. Based on their own mitochondrial DNA work, they suggest our common ancestor evolved around 280,000 years ago—a reasonable, but slightly early, fit with the Out of Africa theory. "It is possible that human individuals—with mitochondrial DNA more divergent from others than known to date—may be found in the future," they say in their paper in the *Journal of Molecular Evolution* in 1991. "However, we think that, because of the extensive collection of the data for individuals from diverse geographic and racial backgrounds, most of the divergence of mitochondrial DNA in the present human populations is accounted for."[22] In addition, it is possible that old mitochondrial DNA lineages may be lost in preference to younger ones, or that one mitochondrial DNA type may have conferred its owners with some slight evolutionary advantage, in which case it would have swept through the human population, disguising its true antiquity. Such notions are possible, but have never been observed in action in animal or human population studies.

Four years after producing the *Journal of Molecular Evolution* paper, Horai produced even more striking backing for the Out of Africa theory, however.[23] He and his colleagues sequenced all 16,500 bases of the mitochondrial genomes of three humans—one each from Africa, Europe, and Japan—as well as four apes: an orangutan, a gorilla, a pygmy, and a common chimpanzee. This was an extraordinarily powerful and complete set of data, and it produced a correspondingly dramatic result. Horai used the ape mitochondrial sequences to get a highly accurate fix on mutation rates among primate populations. Then he applied those rates to the three human lineages and produced a figure that indicated they shared a common ancestor 143,000 years ago. As the African lineage was found to have the most diversity, Horai concluded that the last common ancestor lived there.

The evidence from our organelles may therefore seem conclusive: our cellular power packs were recently made in Africa, and, by inference, so must *Homo sapiens* have been. But there is still that critique of Templeton to consider: that our veneration of mitochondrial DNA blinds us to a broader picture. One little piece of DNA night have a singular source, but that is not necessarily true for the rest of our genome. A gene tree traces back the history of only one fragment of DNA (such as our mitochondrial DNA) but a population tree does not. It, in effect, is an average of many gene trees. So one gene tree may have its roots in Africa, but do the rest? It is a crucial question. So, can geneticists resolve it?

The answer is yes—by studying the other, far more common form of genetic material that we encountered earlier in the chapter: nuclear DNA. This, of course, is made up of tens of thousands of different genes, not just one small piece of DNA. So if we can trace all their roots, we will be able to unravel each gene's ancestry, and should therefore be able to establish without doubt that we have an immediate African pedigree. The trouble is that we inherit our nuclear genes from both parents, in a manner that involves much random shuffling and that makes it impossible for researchers to create exact long lineages with connected branches. This genealogical miasma has not stopped researchers from implementing some highly effective pieces of research, however. Some of this is statistical, but nevertheless illuminating. The rest is highly specific, concentrating, as it does, on particulate pieces of nuclear DNA. Both avenues confirm that we are recent African interlopers, with one of the most dramatic examples of the latter approach being provided by Professors Ken Kidd and Sarah Tishkoff at Yale University's department of genetics.[24]

They have been searching for variations in nuclear DNA that define populations and their relations with one another, and the combination that they have found on chromosome 12 has turned out to be quite extraordinary. The particular targets for their attention were sequences of genetic material called polymorphisms, which often serve no function—like so much of our DNA. Our genes, which direct the manufacture of proteins, are, in fact, a few oases of sense in a desert of nonsense. Over the past few decades, geneticists have found to their surprise that most DNA is junk, long lists of repetitions and meaningless strings of bases uninvolved in protein manufacture. However, these bits and pieces of genetic flotsam often come in several forms. Some of us inherit one kind, others a different type, and this variation can be exploited. In the case of the Yale study, the researchers aimed at two sections that lie close together on chromosome 12 and found that some people lack a long piece of DNA containing 250 of those A, C, G, and T bases. This missing section of genetic material is known as a deletion. Other individuals possess this section, however. And in their other genetic target, people have been found to possess a variable number of repetitions of a little section of five bases, CTTTT. (There are a total of three billion bases that make up the six-foot-length gossamer strands of DNA that are coiled inside a single cell.) Some people have between four and fifteen copies of this little genetic stammer. Again, this variation has no bearing on a person's genetic well-being.

Now, when you look at people living in sub-Saharan Africa, you observe a simple pattern. Individuals have every variety of deletion or nondeletion along with any variety of number of CTTTT repeats. For example, one person may have a chromosome containing the CTTTT sequence repeated eight times as well as a deletion and another person may have a chromosome with the CTTTT sequence repeated twelve times with no deletion. There are many combinations of numbers of repeats and deletions and nondeletions in sub-Saharan Africa. Outside this region—in other words, throughout the rest of the world—you see something very different. Chromosomes with deletions have only one pattern of CTTTT repeats, a sixfold one, while nondeletion chromosomes only have CTTTTs reiterated five or ten times. In other words, Africa shows complete variability. The rest of the world does not. And there is only one feasible explanation: that the small wave of settlers who set off from their African home to conquer the world was made up of a tribe or group of African *Homo sapiens* with only the sixfold CTTTT repetition on their chromosome 12. They carried this combination out to the world 100,000 years ago, and now scientists have picked up its signal like a discarded genetic calling card. (Think of those chromosome 12 variations as a set of genetic dominoes, in which a blank represents the deletion, while the other numbers represent the different repeat possibilities, say a one for a fivefold, two for a sixfold, etc. People in Africa include a mixture of the entire variety of dominoes that have a blank: a blank-zero, a blank-one, etc. However, those in the rest of the world have only one: the blank-two.)

"This says one thing," states Professor Kidd:

It says the rest of the world was peopled from one sub-set of Africans, the ones who had a deletion associated with a six-unit repeat on their chromosome 12, or the ones with a non-deletion and five and tenfold repeat. It also says that only one wave of these people was responsible. And thirdly, it allows us to put a fairly accurate date on that emigration: around 90,000 years ago.

This latter calculation is made by exploiting a basic phenomenon of human genetics—recombination. During the creation of DNA which goes to form the egg and sperm (and future human beings), there is a random shuffling of sections, creating new combinations of genes and DNA sequences. If the association of the chromosome 12 deletion and the six-unit repeat was old, they too would have been reshuffled and would no longer always be linked with each other—as is the situation in Africa. But they are not reshuffled, and to judge from the very few cases where there has been recombination across the globe, fairly simple genetic calculations produce a figure that ranges between 90,000 and 140,000 years for the appearance of these special chromosome 12s, the genetic cargo of the first, and only, wave of modern humans who were then on their way to take over the world. Crucially, this calculation makes no assumptions about DNA mutation rates.

The question is: What does this discovery do for the multiregional hypothesis? "I could be polite about this, and say it poses some serious problems," says Professor Kidd. "If I was being truthfully blunt, however, I would have to say our work blows the theory right out of the water. It is utterly incompatible with the facts that we have uncovered."

Nor is Kidd alone in his emphatic denunciation. There are plenty of other geneticists who have studied nuclear DNA and who read a clear message in its genetic script: that we are young and African in constitution.[25] This work has a lengthy pedigree and goes back to the postwar years when new blood types were being discovered by scientists. They realized that comparisons between different populations' blood group frequencies could reveal how they were related to each other, and could show how long ago any two had separated from a common ancestor race. The greater the similarity, then the closer may be their historical association. For instance, the high frequencies of the B blood group among European Gypsies gave the first real clue that they were Indian in recent origin. Both sets of people have elevated frequencies of the B blood group—about 50 percent, compared with northern Europe, where the figure is only 10 percent.[26]

Similarly, if we look at the distribution of the rhesus negative blood factor we see a clear gradient across Europe. At the eastern edge of the continent, less than 5 percent of the population is rhesus negative. As you head west, its frequency rises until it reaches more than 25 percent on the Atlantic edges of the continent. And that westward increase in rhesus negativity has a simple course. Western European blood (i.e., blood with high frequencies of rhesus negative groups), was probably the normal variety throughout Europe 10,000 years ago. Then, out of the East, came the first farmers, bringing the revolutionary technology of agriculture, and blood that was low in rhesus negative groups. The people of eastern Europe would have been the first to come into contact with these farming newcomers and would have interbred with them, despite, no doubt, some feelings of resentment about their presence. These folk will have absorbed the genes of the first farmers for longer. Rhesus negative blood should therefore be relatively rare among them—as we observe. In the West, we find the opposite is true. Contact came later, so that locals underwent the least genetic mixing with farmers and so they have higher levels of rhesus negative blood—which is what we see. The realization that farming spread across Europe in a westward direction is not new, of course. What is novel and exciting is the knowledge that the blood of these founding fathers and mothers of civilization can be detected in the veins of people today.

Now, the most distinctive of all European "races" are the Basques. They have by far the highest levels of rhesus-negative blood, they speak a language of mysterious origin, and they live in a region, between France and Spain, close to the Lascaux and Altamira caves that provided shelter and artistic wall space for *Homo sapiens* 20,000 years ago. This confluence has led Professor Luca Cavalli-Sforza of Stanford University to the tantalizing conclusion that "Basques are extremely likely to be the most direct descendants of the Cro-Magnon people, among the first modern humans in Europe."

Professor Cavalli-Sforza has been studying the different genetic constitutions of races for two decades, a process which he inaugurated when he began analyzing and differentiating between blood groups (and later other proteins) and their underlying genes. Then he started to draw trees and timetables that tracked the unfolding of our species' racial diversification, a decades-long project that culminated in 1994 in the publication of his massive work, *The History and Geography of Human Genes*, coauthored with Paolo Menozzi and Alberto Piazza.[27] More than 70,000 frequencies of various gene types in nearly 7,000 human population types are included, combined with anatomical, linguistic, and anthropological studies. It is an august body of work that comes down fairly and squarely on the side of the Out of Africa theory. "We conclude a definite preference for the rapid replacement model," states Cavalli-Sforza.[28]

Much of Cavalli-Sforza's study is concerned with genetic distances between populations, a concept that measures the relatedness of one group or tribe with another. "One can estimate degrees of relatedness by subtracting the percentage of rhesus negative individuals among,

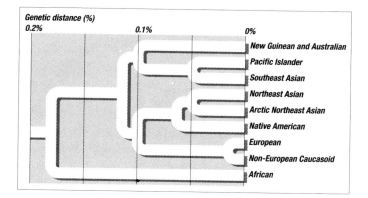

Genetic distance (%)
0.2%   0.1%   0%
                                    New Guinean and Australian
                                    Pacific Islander
                                    Southeast Asian
                                    Northeast Asian
                                    Arctic Northeast Asian
                                    Native American
                                    European
                                    Non-European Caucasoid
                                    African

39 This tree of modern population relationships based on nuclear DNA products
is from the work of Cavalli-Sforza and colleagues. The various African populations have been lumped into a single branch for simplicity.

say, the English (16 percent) from that among the Basques (25 percent) to find a difference of 9 percentage points," states Cavalli-
Sforza. "But between the English and the East Asians it comes to 16
points—a greater distance that perhaps implies a more ancient separation. There is thus nothing formidable in the concept of genetic
distance."

This concept allows scientists to estimate when two populations—
say the English and the Germans—split from their original founding
population and began their own separate existences: "When other
matters are equal, genetic distance increases simply and regularly
over time. The longer two populations are separated, the greater
their genetic distance should be." So Cavalli-Sforza and his collabora-
tors analyzed all the various genetic distances for all the different
peoples of the world, and revealed a picture that is exactly "what
one would expect if the African separation was the first and oldest in
the human family tree." The researchers found that the genetic distance between Africans and non-Africans is roughly twice that between Australians and Asians, and the latter is about twice that
between Europeans and Asians. And these ratios parallel those
produced by the Out of Africa theory: 100,000 years for the separation between Africans and Asians, 50,000 years for that between
Asians and Australians, and 30,000 years for that between Asians and

Europeans. "In these cases, at least, our distances serve as a fair clock," adds Cavalli-Sforza.

More specifically, Cavalli-Sforza looked at certain sequences on two human chromosomes—numbers 13 and 15—and discovered that there was more variety within the African versions of these "microsatellites" than those from other parts of the world. And when he used that diversity to calculate how much time has passed since Africans separated from other populations, he got a figure of 112,000 years. Sounds familiar, doesn't it?

A slightly longer timescale has also been reached by scientists led by Yale's Robert Dorit, who examined a stretch on the Y chromosome, that parcel of DNA that is inherited only through the male line, and which can be viewed as the masculine version of mitochondrial DNA.[29] Dorit and his colleagues compared this sequence in thirty-eight men from different parts of the world and concluded they had a common ancestor, a nuclear Adam, who must have been about 270,000 years old. Other work suggests an even younger age.[30/31]

But inferring patterns of descent from data derived from living people is not unique to the biologist. Just as the genes, and the patterns of genes, which we carried on our African Exodus were modified by the passage of time, so were other human features, ranging from our physiques right down to our speech. Now this last observation provides some unexpected support for the geneticist, and of course, for the Out of Africa theory, because it has become clear that the evolution of the words we speak has close parallels with that of our genes. When a tribe splits, perhaps with one group conquering a new land, the other remaining "at home," the populations so produced accumulate different changes in genes—and language. "People store genes in their gonads and pass them to their children through their genitals; they store grammars in their brains and pass them to their children through their mouths," states Harvard linguistics expert Steve Pinker. "Gonads and brains are attached to each other in bodies so when bodies move, genes and grammars move together."[32]

The process is a slow one, nevertheless, as Steve Jones emphasizes:

In Shakepeare's *As You Like It*, the court jester makes a speech which causes great amusement. Looking at a clock he says "Thus we may see how the world wags; 'tis but an hour ago since it was nine; and after one hour more 'twill be eleven; and so, from hour to hour, we ripe; and then, from hour to hour, we rot and rot; and thereby hangs a tale." Just why this should be so funny is lost on modern audiences—unless they realise that in Shakespeare's time the word "hour" sounded almost the same as the word "whore."[33]

These word "mutations" slowly accumulate so that, very roughly, two languages will lose 20 percent of their common words for every thousand years that their parent population's division have existed separately. An idea of how much a language will change in this period can be gleaned from studying how the Lord's Prayer has mutated over the past ten centuries, a linguistic metamorphosis that can be judged from these samples—outlined in Pinker's book *The Language Instinct*—starting in modern English.

Our Father, who is in heaven, may your name be kept holy. May your kingdom come into being. May your will be followed on earth, just as it is in heaven. Give us this day our food for the day. And forgive us our offences, just as we forgive those who have offended us. And do not bring us to the test. But free us from evil. For the kingdom, the power, and the glory are yours forever. Amen.

Compare this with the King James Bible of 1611:

Our father which art in heaven, hallowed be thy Name. Thy kingdom come. Thy will be done, on earth as it is in heaven. Give us this day our daily bread, and forgive us our trespasses, as we forgive those who trespass against us. And lead us not into temptation, but deliver us from evil. For thine is the kingdom, and the power, and the glory, for ever and ever, amen.

Then read it in Middle English from around 1400 A.D.:

Oure fadir that art in heuenes halowid be thi name, thi kyng-dom come to, be thi wille don in erthe es in heuene, yeue to us this day oure bread ouir other substance, & foryeue to us oure dettis, as we forgeuen to oure dettouris, & lede us not in to temptacion: but delyuer us from yuel, amen.

And finally, contemplate the Old English version, of the period around 1000 A.D.:

Faeder ure thu the eart on heofonum, si thin nama gehalgod. Tobecume thin rice. Gewurthe in willa on eorthan swa swa on heofonum. Urne gedaeghwamlican hlaf syle us to daeg. And forgyf us ure gyltas, swa swa we forgyfath urum gyltedum. And ne gelaed thu us on contnungen ac alys us of yfele. Sothlice.[34]

Clearly languages become incomprehensible with the passage of time until, eventually, one becomes utterly unrecognizable to speakers of the other. For example, English is essentially a Germanic language (despite its borrowings from French, ancient Greek, and Latin). However, the slow trickle of corrupted change has now rendered German quite incomprehensible to the untutored English ear.

So we can see that words accrue infinitesimal changes just like DNA, and, as with our genetic script, we can create a lineage linking each language with its sister tongue. And this is what Cavalli-Sforza did in 1988 when he published a genetic tree of forty-two world populations, together with their respective linguistic affiliations. The speech of the European populations link with those of Indian and Iranian people into the Indo-European language family. This family links, in turn, into the Eurasiatic super-family that includes other groupings containing the languages of the Japanese, Lapps, Eskimos, Siberians, and others. The match between these linguistic branchings and Cavalli-Sforza's gene tree is remarkable. "The tree demonstrates that the genetic clustering of world populations closely matches that of languages," he says. "With very few exceptions, the linguistic families seem to have a relatively recent origin in our genetic tree. In certain cases, a language or family of languages can serve to identify a genetic population."[35] Or as Pinker puts it, modern languages "serve as the fossilised tracks of mass migration in the remote past, clues to how human beings spread out over the earth to end up where we find them today."

Such a confluence, between word and gene, may simply be a coincidence, of course, though that is extremely unlikely, as a team of New Zealand biologists and mathematicians led by Dr. David Penny of Massey University has demonstrated.[36] By using standard computing techniques they have reached what they call an "unambiguous" conclusion. "The two trees are indeed far more similar than expected by chance," they state. There is about a one in 100,000 probability that such similar trees could arise by chance.

This success in creating a bush of Babel has led some scientists to believe they could one day link language with language, language family with language family, until they uncover the Mother Tongue, the first speech uttered by those few global conquerors who emerged on their transcontinental journey 100,000 years ago. Most experts believe this can never be achieved, however. "It is not that I doubt that language evolved only once, one of the assumptions behind the search for the ultimate mother tongue," states Steve Pinker:

> It is just that one can trace words back only so far. Most linguists believe that after 10,000 years, no traces of a language remain in its descendants. This makes it extremely doubtful that anyone will find extant traces of the most recent ancestor of all contemporary languages, or that that ancestor would in turn retain traces of the language of the first modern humans who lived 200,000 years ago.

And there we have it. The blood that courses through our veins, the genes that lie within our cells, the DNA strands that nestle inside our mitochondrial organelles, even the words we speak—all bear testimony to the fact that 100,000 years ago a portion of our species emerged from its African homeland and began its trek to world dominion. (The other part, which stayed behind, was equally successful in diversifying across the huge African continent, of course.) It may seem an exotic, possibly unsettling, tale. Yet there is nothing strange about it. This process of rapid radiation is how species spread. The real difference is just how far we took this process—to the ends of the earth. A species normally evolves in a local ecology that, in some cases, provides a fortuitously fertile ground for honing a capacity for survival. Armed with these newly acquired anatomies, or behavior patterns, it can then take over the niches of other creatures. This is the normal course of evolution. What is abnormal is the supposed evolution of mankind as described by the multiregionalists. They place their faith in a vast global genetic link-up and compare our evolution to individuals paddling in separate corners of a pool, as we have seen. According to this scheme, each person maintains their individuality over time. Nevertheless, they influence one another with the spreading ripples they raise—which are the equivalent of genes flowing between populations.

Let us recall the words of Alan Thorne and Milford Wolpoff, quoted in Chapter 3. They state that:

> The dramatic genetic similarities across the entire human race do not reflect a recent common ancestry for all living people. They show the consequences of linkages between people that extend to when our ancestors first populated the Old World, more than a million years ago. They are the results of an ancient history of population connections and mate exchanges that has characterised the human race since its inception. Human evolution happened everywhere because every area was always part of the whole.[37]

Gene flow is therefore crucial to the idea that modern humans evolved separately, for lengthy periods, in different corners of the earth, converging somehow into a now highly homogeneous form. Indeed, the theory cannot survive without this concept—for a simple reason. Evolution is random in action and that means that similar environmental pressures—be they associated with climate change, or disease, or other factors—often generate different genetic responses in separate regions. Consider malaria, a relatively new disease that spread as human populations became more and more dense after the birth of agriculture. Our bodies have generated a profusion of genetic ripostes for protection in the form of a multitude of partially effective inherited blood conditions. And each is unique to the locale in which it arose. In other words, separate areas produced separate DNA reactions. There has been no global human response to malaria.

Nevertheless, multiregionalism maintains that gene flow produces just such a global response. Given enough time gene exchange from neighboring peoples will make an impact, its proponents insist. This phenomenon, they say, has ensured that the world's population has headed towards the same general evolutionary goal, *Homo sapiens*; though it is also claimed that local selective pressures would have produced some distinctive regional physical differences (such as the European's big nose). And if the new dating of early *Homo erectus* in Java is to be believed (as many scientists are prepared to), then we must accept that this web of ancient lineages has been interacting— like some ancient, creaking international telephone exchange—for almost two million years.

Now this is an interesting notion which makes several other key assumptions: that there were enough humans alive at any time in the Old World over that period to sustain interbreeding and to maintain the give and take of genes; that there were no consistent geographical barriers to this mating urge; that the different human groups or even species that existed then would have wished to have shared their genes with one another; and that this rosy vision of different hominids evolving globally towards the same happy goal has some biological precedent.

So let us examine each supposition briefly, starting with the critical question of population density. According to the multiregionalists, genes had to be passed back and forward between the loins of ancient hominids, from South Africa to Indonesia. And this was done, not by rapacious, visiting males spreading their genotype deep into the heart of other species or peoples (a sort of backdoor man school of evolution), but by local interchange. In neighboring groups, most people would have stayed where they were, while some individuals moved back and forward, or on to the next group as they intermarried. In other words, populations essentially sat still while genes passed through them. But this exchange requires sufficient numbers of neighboring men and women to be breeding in the first place. By any standard, hominids—until very recent times—were very thin on the ground. One calculation by Alan Rogers, a geneticist at the University of Utah, in Salt Lake City, and colleagues uses mitochondrial DNA mutations to assess how many females the species possessed as it evolved. The results he produced are striking. "The multiregional model implies that modern humans evolved in a population that spanned several continents, yet the present results imply that this population contained fewer than 7,000 females," he states in *Current Anthropology*.[38] It is therefore implausible, he adds, that a species so thinly spread could have spanned three continents and still have been connected by gene flow.

Then there is the question of geography. To connect humanity throughout the Old World, genes would have had to flow ("fly" might be a better word) back and forth up the entire African continent, across Arabia, over India, and down through Malaysia; contact would therefore have had to have been made through areas of low population density such as mountains and deserts, coupled with some of the worst climatic disruptions recorded in our planet's recent geological past. Over the past 500,000 years, the world was gripped by frequent Ice Ages: giant glaciers would have straddled the Himalayas, Alps, Caucasus Mountains; meltwater would have poured off these ice caps in torrents, swelling inland lakes and seas (such as the Caspian) far beyond their present sizes; while deserts, battered by dust storms, would have spread over larger and larger areas. Vast regions would have been virtually blocked to the passage of humans. At times our planet was extremely inhospitable while these straggling hordes of humans were supposed to be keeping up the very busy business of cozy genetic interaction. "Even under ecologically identical conditions, which is rarely the case in nature, geographically isolated populations will diverge away from each other and eventually become reproductively isolated. . . . It is highly improbable that evolution would take identical paths in this multi-dimensional landscape," writes the Iranian researcher Shahin Rouhani.[39]

Cavalli-Sforza agrees: "What is very difficult to conceive is a parallel evolution over such a vast expanse of land, with the limited genetic exchange that there could have been in earlier times."[40] He acknowledges that it is theoretically possible that the genes of west European humans would have been compatible with those of east Asia despite their ancient separation. Barriers to fertility are usually slow to develop: perhaps a million years or more in mammals. However, he adds, "barriers to fertility of a cultural and social nature may be more important than biological ones." Two very different looking sets of people may have been able to interbreed physically but would have considered such action as breaking a gross taboo.

In other words, we are expected to believe that a wafer-thin population of hominids, trudging across continents gripped by Ice Ages, were supposed to be ready to mate with people they would have found extraordinarily odd-looking and who behaved in peculiar ways. Cavalli-Sforza, for one, does not buy this. "Proponents of the multiregional model simply do not understand population genetics," he states. "They use a model that requires continuous exchange of genes, but it requires enormous amounts of time to reach equilibrium. There has been insufficient time in human history to reach that equilibrium." The spread of modern humans over a large fraction of the earth's surface is more in tune with a specific expansion from a nuclear area of origin, he adds.

Now this last point is an important one, for it is frequently presented in the popular press that the Out of Africa theory represents a divergence from the natural flow of biological affairs, that its protagonists are somehow on the fringes of orthodoxy, proposing strange and radical notions. The reverse is true—the large number of scientists quoted in this chapter indicates the wide intellectual support now accorded the theory. It is a very new idea, admittedly. It is only a little more than a decade since it was first proposed, on the basis of fossils, by scientists like Bräuer and Stringer. Yet its precepts now affect many areas of science, and its implications are accepted by their most distinguished practitioners. We are witnessing a rare moment in science, the replacement of a redundant orthodoxy by a formerly heretical vision. Hence the words of Yoel Rak as he staggered from a multiregionalists' symposium in 1991. "I feel like I have just had to sit through a meeting of the Flat Earth Society," he moaned.[41]

Of course, Rak became an African Exodus proselytizer many years ago. A more damning convert, if you are multiregionalist, is that of *Science*, a journal noted for its dispassion and conservatism. "The theory that all modern humans originated in Africa is looking more and more convincing," it announced in March 1995, "and the date of the first human exodus keeps creeping closer to the present . . . the evidence coming out of our genes seems to be sweeping the field."[42]

In fact, the idea that the opposition—the multiregionalists—represent the norm in biological thinking is to present the story of human origins "ass backwards," as Stephen Jay Gould succinctly puts it.[43]

> Multi-regionalism . . . is awfully hard to fathom. Why should populations throughout the world, presumably living in different environments under varying regimes of natural selection, all be moving on the same evolutionary pathway? Besides, most large, successful, and widespread species are stable for most of their history, and do not change in any substantial directional sense at all. For non-human species, we never interpret global distribution as entailing preference for a multiregional view of origins. We have no multiregional theory for the origin of rats or pigeons, two species that match our success and geographical spread. No one envisions proto-rats on all continents evolving together toward improved ratitude. Rather we assume that *Rattus rattus* and *Columbia livia* initially arose in a single place, as an entity or isolated population, and then spread out, eventually to cover the globe. Why uniquely for humans, do we develop a multiregional theory and then even declare it

orthodox, in opposition to all standard views about how evolution occurs?

The answer to that critical question has much to do with an outlook that has pervaded and bedeviled science throughout history. We have, at various times, been forced to abandon species-centric scientific notions that we live at the center of the cosmos, and that we were specially created by a supreme being. A last vestige of this urge to self-importance can be seen in multiregionalism, which holds that our brain development is an event of all-consuming global consequence towards which humanity strived in unison for nearly two million years. It argues that *Homo sapiens*' emergence was dictated by a worldwide tendency to evolve large braincases, and share genes and "progress." Humanity is the product of a predictable proclivity for smartness, in other words, so we cannot possibly be the outcome of some local biological struggle. Surely that would demean us. To believe that humanity could be the product of a small, rapidly evolving African population who struck it lucky in the evolution

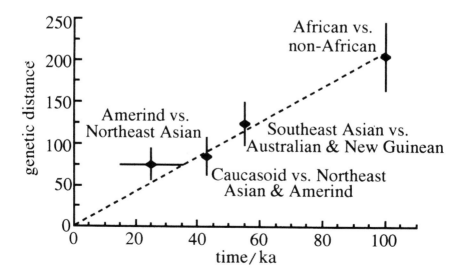

40 Joanna Mountain and Cavalli-Sforza compared genetic distances between modern peoples with archeological and fossil evidence of their separations. They match well over a timescale of 100,000 years but would not fit much longer divergence times.

stakes is therefore viewed as being worse than apostasy by these people. Unfortunately for them, there is little proof to support their specialist, global promotion of mankind—as we have seen. Once again we must adopt the simplest scientific explanation (i.e., the one for which the facts best fit) as the superior one. As this chapter has made clear, there is no good genetic evidence to sustain an argument that places humanity on a plinth of global superiority. To do so is to indulge in mysticism. *Homo sapiens* is not the child of an entire planet, but a creature, like any other, that has its roots in one place and period—in this case with a small group of Africans for whom "time and chance" has only just arrived. Nor is our species diminished in any way by such interpretations. Indeed, we are enriched through explanations that demonstrate our humble origins, for they place us in an appropriate context that, for the first time, permits proper self-evaluation and provides an understanding of the gulf we are crossing from a clever ape to a hominid that can shape a planet to its requirements—if only it could work out what these are.

It is now time to look at our arrival on the planetary stage, its timetable, the nature of the people who made it, and the extraordinary tools employed by science in revealing this startling story.

6

Footprints on the Sands of Time

Lives of great men all remind us, We can make our lives sublime,
And, departing, leave behind us, Footprints on the sands of time.

Henry Wadsworth Longfellow

Our African Exodus was the greatest of all human journeys, a global endeavor that took our ancestors over every conceivable obstacle thrown up by nature: estuaries, deserts, mountain ranges, steppes and tundra, dense forests, fields of ice and snow, and sheer distance—the 9,000-mile length of the Americas being a prime example. It is a testimony to human resilience and resourcefulness that we overcame these hurdles in a few dozen millennia, leaving only a handful of isolated, ocean islands and the polar caps unconquered until recent times.

And of all the impediments faced by our African antecedents, the most severe would have been the open seas, though in some cases, growing ice caps may have stolen enough water to create land bridges that have long since disappeared. During the Ice Age, "Doggerland" (now the Dogger Bank under the North Sea) linked Britain and Europe, for example, and trawlermen still occasionally drag up mammoth and woolly rhino bones from this drowned world. Similarly, Beringia (now the Bering Straits) joined Siberia and Alaska, and allowed humans to launch their onslaught on the Americas. Other sea barriers were less accommodating, however—like those between North Africa and Gibraltar, and between Java and Australia. Despite the locking up of much of the world's sea into ice, these waters never parted to permit the passage of human beings. Much nautical ingenuity (with a fair smattering of good luck) was needed to cross these obstacles.

On top of such geographical headaches, there was the simple, but inconvenient matter of other human species. Neanderthals were entrenched in Europe until relatively recently, while on the other side of the world, the Ngandong folk persisted in Java, along with the descendants of China's Dali people, and may have been equally stubborn and reluctant to make way for modern humans.[1] They too were supplanted, although unlike in Europe, the details of the process are, for the moment, completely lacking—we only see its aftermath.

We triumphed in the end for a variety of reasons: social, cognitive, behavioral, and technological. In this last category, implements and creative techniques helped take our ancestors to nooks and crannies that would have been barred to them if they had begun their dispersal unaided. The remains of most of these tools—made of hides and wood—have long rotted away, of course. However, to judge from residual stone implements, and from the behavior of hunter-gatherers today, we can deduce they probably had warm clothing sewed together using carved bone and antler needles, water containers made from hides, boats and rafts made from fallen tree trunks or bamboo tied together, sophisticated foraging techniques, and the use of fire and smoke to burn clearings and trap prey. In these ways our ancestors were able to open up previously uninhabited lands, and conquer ones that were already peopled.

Today, as modern science tries to trace that ancient odyssey, we face a different sort of obstruction—the barrier of time. We can pinpoint promising sites where our predecessors camped or hunted on their journey, but how do we put the resulting discoveries into context? How do we fix a chronology for their progress round the globe? Small lumps of bone and stone tend to be fairly uncommunicative on this matter, unless they are treated with skill and expertise. Fortunately, modern science has equipped us with a veritable arsenal of weapons to unravel these twisted strands of human prehistory. This chapter will look at some of the most exciting of these and show how dramatically each has contributed to our new understanding of humanity's recent emergence from its African homeland.

One of the first, and most important, technologies to be developed has been that of radiocarbon dating, which exploits the fact that plants constantly take in carbon dioxide and absorb its carbon. This is then passed on to animals when they eat the plants. Some of this element contains a naturally occurring radioactive isotope that decays slowly, and so, by measuring how much radiocarbon a piece of plant, charcoal, or bone contains, a reasonable estimate of its age can be obtained. This technique has become one of the most important weapons in the archeologists' armory, although—as we shall see—it has its limitations. Developed from research that led to the atomic bomb, American scientists first used it in 1949 to produce dates for several ancient Egyptian sites. These fitted well with previous age estimates of the Pharaohs' dynasties, and the technique quickly acquired respectability. Since then it has notched up several triumphs:

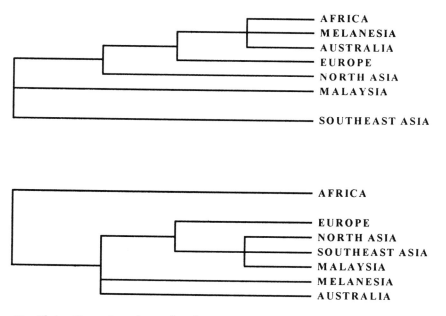

41   Christy Turner's analyses of tooth variations suggest that modern Southeast Asians are closest to our ancestral condition (above). However, analyses using fossils suggest instead that shared features in African and Australian teeth may be ancestral characters (below).

Stonehenge was shown to have been built a thousand years earlier than previously believed, while the Agricultural Revolution—when crops were grown and animals tamed for the first time—was demonstrated to have originated in the eastern Mediterranean at least 10,000 years ago, double some previous estimates of its age.[2]

However, radiocarbon dating's greatest impact on the public imagination came with a far more controversial story: when it helped to unravel the secrets of a skull found in a gravel pit at Piltdown in Sussex that had an apelike lower jaw and modern-looking braincase.[3] This object was declared to be a genuine "missing link" in 1912 by scientists (especially those in the U.K., who were anxious to find a British equivalent to all those strange and important fossils that were being found in France, Germany, and the Dutch colony of Java). The trouble was that, as other ancient hominid remains were uncovered in Europe, Asia, and Africa, none proved to be remotely like the Piltdown skull. Scientific suspicion mounted, until in 1953, stringent tests showed it was probably a fake, an amalgam of human braincase and orangutan jaw, chemically stained to give it an ancient appearance. So scientists turned to radiocarbon dating for confirmation, and discovered that the human and ape bones were only a few hundred years old, corroborating the view that the Piltdown skull was a fraud. (The perpetrator remains a mystery, however, with suggestions about his identity ranging from Charles Dawson, the solicitor who found the first bits of skull, to the distinguished anthropologist Sir Arthur Keith, and even Sherlock Holmes's creator, Sir Arthur Conan Doyle.) More recently, radiocarbon dating has demonstrated that fibers from the supposed cerement of Christ (the Turin Shroud) are only about 700 years old, showing it is a medieval forgery.

Despite radiocarbon dating's enormous impact on our understanding of this planet's recent prehistory, the technology has some decidedly unworldly roots, however. The unstable carbon isotopes that it exploits originate in outer space. Many miles above our heads, cosmic rays—beams of high energy particles that emanate from outside our solar system, and possibly our galaxy—strike atoms of nitrogen, the chief component of our atmosphere, transforming them into an isotope of carbon called carbon-14. This is radiocarbon and it is chemically identical with normal carbon. Both are continually absorbed into the bodies of living things. That process stops when an organism dies, and its store of radiocarbon begins to decay, back into atoms of nitrogen. After about 5,700 years, half is left; after about 11,400 years a quarter; and after about 17,000 years, only an eighth. In this way, the date of a piece of material can be calculated by measuring how much its radiocarbon has decayed compared with its normal carbon content (which does not fluctuate over time).

However, there is clearly a limit to how far we can exploit these diminishing fractions in order to peer back into our prehistory. After about 35,000 years, less than 2 percent of a sample's original radiocarbon will remain. In addition, material may have been poorly preserved, so that its carbon will have been lost, if, for example, it was buried in acidic soil. Even worse, any slight contamination with carbon from another source has a severe effect when dating very old materials. For example, a 35,000-year-old sample would appear to be about 4,000 years younger than it really is if an impurity of only 1 percent of new carbon was added to it. As a result, most scientists now treat radiocarbon ages of more than about 30,000 years with considerable caution.

Unfortunately, for the decades that followed 1950, researchers had to make do with this limited acuity when looking into the past. There was radiocarbon dating, and that was it, which made the study of our recent African origins, and the spread of modern people between 50,000 to 100,000 years ago, a tricky business. African sites older than a million years were often embedded in volcanic rocks which could be accurately dated using a technology called potassium-argon chronology. However, when it was first developed, this technique was only practical for dealing with sites older than 500,000 years. In any case, Europe and the Middle East had no such convenient volcanoes, and so scientists could only twiddle their radiation counters. It was an awkward problem. Then, in the 1980s, a new generation of techniques began to make an impact: electron spin resonance, uranium-series dating, and two forms of luminescence dating. (None roll off the tongue with any ease at all, as James Shreeve, writing in *Discover*, acknowledges. As he points out: "The terrain of geochronology is full of terms long enough to tie between two trees and trip over."⁴) Each would play a significant role in revolutionizing knowledge of our prehistory.

Take the example of luminescence dating. Its basic principles are actually quite old, having been discovered by Robert Boyle in 1663, as part of one of science's oddest cases of serendipity. Boyle had borrowed a diamond and had taken it to bed with him. Unfortunately, history does not record why the distinguished British physicist felt the urge to sleep with a loaned gem. Nevertheless, science should be grateful. Boyle found that when he rested the jewel "upon a warm part of my naked body" it gave off a glow. (The mind boggles at what he was doing.) So impressed was Boyle with this phenomenon that he presented a paper about his discovery to the Royal Society—the next day.

The luminescence signal that Boyle had discovered is produced when natural radiation damage, built up over time in materials like diamond, sand, or flint, is released. These subatomic blemishes occur because radiation knocks electrons out of their correct orbits around atoms, and some get trapped in areas of impurity in crystalline substances. Only when the sample receives further energy, for instance in the form of heat, are they released, giving off light as they do so. The more luminescence there is, the further back the buildup of those errant electrons must have begun. However, it is important to note that these electron clocks are constantly being reset to zero—for instance when sand grains are bleached in the sun, when a piece of pottery is baked, or when a tool is burned in an ancient campfire. Crystals in stones or cooking implements then give off their trapped electrons, after which they start to acquire new ones. The point is that through the controlled liberation of those electrons, acquired since that fiery immersion, the age of a flint or a clay hearth can be calculated. This is done either by heating (thermoluminescence) or by using a laser (optically stimulated luminescence) to release a burst of illumination that can be measured with a device known as a photomultiplier. The bigger the spike of light, then the more trapped electrons must lie within the sample's crystal lattices, and the greater the elapsed time since its last exposure to heat. (The related method, electron spin resonance, uses microwave radiation to count electrons within their tooth enamel crystalline traps and arrive at the same result.) In this way, the age of once-burned flint or shard of a pot, baked in an ancient oven, can be calculated by one of those fire-maker's descendants in the pristine sophistication of a modern laboratory. "Luminescence has revolutionised the whole period I work in," says archeologist Rhys Jones, of the Australian National University. "In effect, we have at our disposal a new machine—a time machine."[5]

This leaves us with the last of our technological tetralogy, uranium-series dating, used mainly when working on cave deposits, and which exploits the fact that uranium has several naturally occurring radioactive isotopes. These are laid down within a stalagmite as it forms, for example. And as these radioactive elements decay, they produce daughter products. By measuring the ratio of parent element to atomic offspring, the age of a deposit can be calculated.

The precise power of luminescence, electron spin resonance, and uranium series has allowed scientists to reach well beyond the limits of radiocarbon dating to far more distant times. In some cases, our new prowess has confirmed previous ideas—for example, that Neanderthals and Cro-Magnons coexisted in Europe 30,000 to 40,000 years ago. In other cases—for instance in the Middle East—they have turned established views topsy-turvy. We have already seen that it had been assumed Neanderthals had arrived in the Levant first (at sites like Tabun, Amud, and Kebara), to be followed by early moderns (who lived at Skhul and Qafzeh). But when scientists focused their luminescence detectors on burned flints (scorched when dropped in campsite fires, it is presumed), and their electron spin resonance techniques on the teeth of contemporary animals, they found Neanderthals had been living at Tabun 110,000 years ago, and at Kebara and Amud 50–60,000 years ago, while, in between, early moderns were clearly thriving at Skhul and Qafzeh around 100,000 years ago. It was one of the most critical discoveries to have emerged in our Out of Africa saga.

In addition, these techniques helped reverse the view that Africa was a backward Stone Age ghetto. Stone studies showed that cultural life may have been running ahead of Europe and the Middle East between 50–100,000 years ago, while Africa's bone record also revealed surprises. For example, electron spin resonance demonstrated that the Jebel Irhoud fossils from Morocco (see Chapter 1) were probably three times older than the previous 50,000-year-old estimate. Other analyses have shown that a near modern human skull found at Ngaloba in Tanzania was at least 130,000 years old, while one at Singa in the Sudan was 150,000 years old. These African hominids were living in the right place and right time to be the real ancestors of the first modern people.

But if there were primitive modern humans in Africa 100,000 years ago, why did they take so long to reach Europe, Asia, Australia, and the Americas? For instance, in Asia, there is little evidence of any *Homo sapiens* prevalence (apart from in the Levant) until about 40,000 years ago. We catch glimpses of their presence at contemporary sites like K'sar Akil in Lebanon and Darra-i-Kur in Afghanistan; and then in Sri Lanka about 30,000 years ago; in China, about 25,000 years ago; and in Japan about 17,000 years before present.

So what happened in between? Where were our ancestors lurking and what were they doing? Puzzlingly, there are few good answers to these questions. Indeed, we have greater knowledge about more distant periods of our prehistory than this crucial, recent era: paleontologists studying this period would kill for a dated skeleton as well preserved as the 1.5 million-year-old remains of the Nariokotome boy. All we can say is that in Africa, the archeological record tells us that people were certainly living there between 40,000 and 80,000 years ago, although the associated fossils are disputed and scrappy. To find more information about this vital prehistoric time we therefore have to seek clues elsewhere, not from bones, but from genes which we know can be every bit as informative as fossils. As we have seen, each set of human genes has a different history. Some, like the ABO blood groups, we share with chimpanzees and gorillas, indicating they must have been part of our biological heritage in the last five million years. Others have differentiated greatly since that time, like mitochondrial DNA. To exploit these varying features, researchers try to combine information to illuminate past population fluctuations, for instance to see if there were filters in our history which sucked off variation. These events—called bottlenecks—would have occurred when numbers crashed because of drought, volcanic eruptions, or other natural calamities. People perished, slicing their genetic individuality from posterity.

Just consider our mitochondrial DNA's remarkable uniformity, a certain sign of a recent bottleneck, say scientists. As they point out, if we are all so alike today, then there could only have been a very few of us sharing this diminished pool of mitochondrial genes a short time ago. Now it is generally believed this numerical compression occurred about 100,000 years ago when *Homo sapiens* had only about 10,000 adult members. This calculation is supported by work on nuclear DNA, a much more varied form of genetic material that also contains data on older genetic diversities. These permit scientists to peer through the neck of our population bottle into its historic recesses, where they see signs that there were at least 100,000 adult archaic forebears of our African ancestors about 200,000 years ago. Human numbers, although small by present-day standards, would have been big enough to disseminate over Africa, Asia, and Europe. But later, there was either a population crash or our ancestors became isolated from the rest of humanity. As a result, the newly-born population of modern humans was reduced to only 10,000 adults. We would have been too thinly spread for colonization, and would have been forced to cling to only a small stretch of territory, leaving the rest of the world—for the time being—to the Neanderthals and their kin. Not every scientist agrees that this drop in numbers would necessarily have been that threatening a reduction. Others, such as Henry Harpending, argue the contrary. "Our ancestors survived an episode where they were as endangered as pygmy chimpanzees or mountain gorillas are today," he says.[6]

What is clear is that our genes reveal a fleeting image of the rise and fall (and rise again) of *Homo sapiens*. From a widespread ancestral population, stable in numbers, we suddenly plunged towards biological insignificance. Then, according to scientists who have compared particular genes in individuals from the same and from other populations, we bounced back. They have plotted the number of mutational differences between samples, producing a shape like a hill or series of hills with the biggest peak showing where and when most differences are concentrated.[7] Mutations happen all the time, of course, but if a population suddenly grows in size, those that have just occurred will be copied disproportionately, leaving an unmistakable genetic footprint. Twenty-six groups of people, including Bushmen, Sardinians, New Guineans, and the Nuu Chah Nulth of North America, were studied this way, and twenty-four revealed expansion peaks between 40,000 and 80,000 years ago. (The two exceptions being a pair of African populations that seem to have suffered a much more recent bottleneck.) The decisive point is that each of these two dozen evolutionary phoenixes rose in relative isolation from each other. In other words, *Homo sapiens* went through a near fatal numerical crash or isolation between 50,000 and 150,000 years ago and then bounced back at different times, rates, and places. Our African recovery seems to have begun first, perhaps 60,000 years ago, followed by Asia 50,000 years ago, and finally by peripheral areas such as Europe and Australia, around 40,000 years ago. Each of these twenty-four populations was already separated when their numbers increased, however. And that leaves us with two very awkward questions. What caused the bottleneck in the first place? And what drove the growth in their separate populations?

Now these are vitally important questions, whose answers will clearly show that time and chance—and not predestination to greatness—played a pivotal role in our emergence as global conquerors. Certainly in the case of our first query, it seems obvious once again that that mighty equalizer—climatic change, encountered so many times before in this book—played a dramatic role. About 150,000 years ago, a 60,000-year-old cold "snap" was peaking. Ice caps sprawled across the poles, bringing cooler, drier conditions to the rest of the planet. The Sahara Desert had expanded, virtually cutting off North Africa from the rest of the continent, while the Kalahari Desert swelled across the south, forming a second, almost impenetrable band. At the same time, central Africa's dense tropical forests shrank into separate western and eastern refuges, surrounded by grasslands that would have provided homes for humans. It may have been south of the then impenetrable Sahara that our species was forged. Then about 130,000 years ago, the climate switched back briefly into a warmer, moister mode. The deserts began to retreat and the forests to expand again, a situation that probably led to prototype modern humans' first tentative steps out of Africa into the Middle East 120,000 years ago and further into Asia by 80,000 years ago.

These intercontinental intruders were the first unequivocal representatives of *Homo sapiens* and they must have evolved in Africa's hinterland between 130,000 to 200,000 years ago—during the long spell of global cooling—from archaic human predecessors. These hominid newcomers spread and by 100,000 years ago had established themselves in terrain stretching from southern Africa to present-day Ethiopia and the Levant. But where did they come from originally? The remains of these people, and evidence of their behavior, is tantalizing sparse, although we suspect that once again East Africa may turn out to be the key to our origins. However, we desperately need more evidence from across the continent to confirm this idea. What we do know is that this transition turned people with rather broad, long, and low braincases, with quite strong browridges (like the Florisbad and Jebel Irhoud remains), into individuals with higher, shorter, and narrower crania with smoother foreheads (like the Kibish or Border Cave fossils).[8] Chins became a permanent facial feature even in children, though the rest of the face remained short, broad, and flat, with a wide nose and low, well-spaced eye sockets—which, in life, might have enclosed brown irises with epicanthic folds around them. The skeleton retained its lanky tropical shape, but bone thickness and muscle power was decreasing.

The reason for this physical metamorphosis is also a puzzle, though the reduction of skeletal strength gives an important clue—that our ancestors were developing a more energy efficient lifestyle, with brain predominating over brawn for the first time in human evolution. We shall look at what these behavioral changes might have been in Chapter 8, though it is unclear whether the driving force for our transformation was a change in our brains, societies, or technologies. All we can say is that isolation and stress in those cold and dry days, about 150,000 years ago, were probably the triggers for this fundamental change in humanity.

Not that the hard times came to an end once this fledgling hominid species emerged from Africa and tried to make its way in the world. The earth was gripped by continuing climatic mayhem as changes in its orbit began inexorably to push down the world's thermostat. Then to add to these woes, about 74,000 years ago, Mount Toba on the island of Sumatra exploded in the largest volcanic eruption of the past 450 million years. The blast was 4,000 times more powerful than that of Mount St. Helens and would have sent more than 1,000 cubic kilometers of dust and ash into the atmosphere, plunging the earth into years-long volcanic winters.[9] Summer temperatures could have dropped by as much as twelve degrees centigrade, while forests shrank, deserts spread, and in eastern Asia, a prolonged winter monsoon would have swept clouds of dust from inland deserts round the globe. This, says Stanley Ambrose of the University of Illinois, could have been the cause of *Homo sapiens'* population crash.[10] Having evolved in warm savanna sun, we nearly perished, huddled in cold, dismal misery as volcanic plumes straddled the earth. The links between modern human pioneers in Asia and their motherland were severed, allowing cold-adapted Neanderthals to become the sole occupants of the Middle East for the next 30,000 years.

Ironically all this fragmentation and environmental pressure may have been the stimulus for the final crucial changes that transformed these hominid bit players into masters of the planet. Forged in this bleak crucible, evolutionary pressures triggered alterations to our brains and social behavior and we were sent "ticking like a fat, gold watch" towards zoological stardom. Some scientists say they can already detect signs of innovations associated with these changes at Klasies, Border Cave, and Katanda around 100,000 years ago.[11] There they have found remains of the use of that Upper Paleolithic perennial, red ocher, along with the signs of complex composite tools of wood and stone. However, other researchers believe these innovations appeared later, about 50,000 years ago, nearer the time of our big rebound from near extinction.[12]

Of course, there was clearly no single exodus, no one triumphant army of early hunter-gatherers who were led Out of Africa toward a new world by a Paleolithic Moses. Instead, our exodus would have occurred in trickles as our ancestors slowly seeped out of the continent, expanding their hunting ranges and taking over new territory. Marta Lahr and Robert Foley of Cambridge University believe they can reconstruct one such expansion that spread eastward out of the Horn of Africa about 80,000 years ago.[13] Its populations diversified as they moved to eastern and southeastern Asia, forming the region's modern "races." A later dispersal, about 50,000 years ago, infiltrated North Africa, western Asia—and Europe, in the form of our old friends, the Cro-Magnons.

Now in the past, it has been assumed that these first African migrants must have been black, like so many people on the continent today. Only later did some members evolve lighter, white and brown skins, it was thought. However, Jonathan Kingdon has attempted a detailed reconstruction of early human dispersals in his book *Self-Made Man and His Undoing* and concludes that our original skin color was probably a medium brown. According to him it was only later, as early modern humans moved along Asia's southern seaboard, settling along the coast as they progressed, that their appearance changed. They became dependent for food on a life that was governed by the sea. Selection would therefore have favored those who could stay out in the sun, those of the darkest color, when the tides required it. Hence black skin evolved for the first time, and the genes responsible began to spread across southern Asia, with some ending back in mankind's African homeland. "On present evidence, modern humans are likely to have begun with all the built-in advantages of a versatile light brown skin and only developed later the extremes of densely shielded (black) or totally depigmented skins," says Kingdon.[14]

But to what degree did these ancient members of *Homo sapiens* look like people alive today? It is, after all, a specific prediction of the Out of Africa theory that racial characteristics are new and relatively unimportant facets of our species' anatomy. So can we detect evidence in skeletons to support this idea? Intriguingly enough, when we examine some of the oldest *Homo sapiens* relics, like those

100,000-year-old fossils from Qafzeh and Skhul, we find they do not show the kinds of differentiation that distinguish races today. Their skeletons are modern, as is the overall shape of their braincases, but they have unusually short, broad faces, with short, wide noses. Nor does the picture get any clearer when we move on to the Cro-Magnons, the presumed ancestors of modern Europeans. Some were more like present-day Australians or Africans, judged by objective anatomical categorizations, as is the case with some early modern skulls from the Upper Cave at Zhoukoudian in China.[15] It is a confused picture and suggests that racial differences were still developing even relatively recently, and should be viewed as a very new part of the human condition. It is an important point, for it shows that humanity's modern African origin does not imply derivation from people like current Africans, because these populations must have also changed through the impact of evolution over the past 100,000 years.

So far we have seen that genes, bones—even electrons trapped in crystals of tooth enamel and stone—can reveal unguessed secrets about our prehistory, although this is not the only data we can study. There is the example of dental evidence, for example. The exact shape of our teeth is under genetic control, and dissimilar populations display different patterns of growth and form. Many Oriental populations have a "scooped out" or "shoveled" form on the inner surface of their upper front teeth (incisors); Africans have a high frequency of a seventh cusp on their first lower molars; Europeans typically have four cusps on their lower second molars; and Australians usually have three roots on their upper second molars. High or low frequencies of these characters can be used to separate or link different populations.[16] For example, the fact that many Native American people have shovel-shaped incisors like East Asians provides powerful support for the idea that Asia was once their homeland.

It all seems to paint a fairly fragmentary but nonetheless cogent picture of our exodus from Africa, one that suggests a fundamental timetable for our colonization of the world. Firstly, we moved from Africa to Asia about 100,000 years ago, and spread eastward until we reached New Guinea and Australia by about 50,000 years ago. A little later, having conquered the East, mankind also dispersed westward from Asia and drifted into Europe, eventually extinguishing Neanderthals there. Finally, at some point, Asian people made their way over Beringia and speedily down through the Americas, their progress unencumbered by the presence of other hominid competitors.

By 30,000 years ago, modern humans had achieved an estimated breeding population of at least 300,000 individuals. We were then the only human species left on earth, probably the first time the bush of human evolution had been pruned to a single branch for more than a million years. The others had withered in the face of repeated cold shocks between 75,000 and 30,000 years ago. These plunged the oceans and then continents into a series of mini–Ice Ages, each lasting one or two millennia.[17] Humanity's ailing non-*sapiens* branches must have suffered a slow attrition of numbers in the face of such climatic instability and as a result of more adaptable *Homo sapiens'* faster growing populations. The descendants of Java's Ngandong and China's Dali people may have gone under first, while those remarkable survivors, the Neanderthals, clung on in shrinking pockets, such as Zafarraya, until 30,000 years ago.

It is a neat image, carefully crafted from the work of scientists who have studied Ice Age sediments, probed our DNA, and honed the technologies of radiocarbon dating, luminescence, and all their other wizardry to a state in which we can begin to see the shadowy footprints left behind by our predecessors on their African Exodus. Their detective work is a triumph of modern science, but we should not run away with the idea that it has left us with a crystal-clear vision of our past. There are many puzzles still to be resolved, of which there are two particularly vivid, if not to say glaring, examples: the peopling of the Americas and Australia. So let us conclude this chapter with a look at these two vast lands, and investigate their mysterious and controversial position in the story of our origins.

As we have seen, evidence from our teeth—and other sources—provides a clear indication that Asian tribes crossed into Alaska before heading down through the entire 9,000-mile length of North

and South America via the lost land of Beringia. This was a colonization and a human challenge on a vast scale, taking humanity from the
chill tundra of the Arctic Circle to the near Antarctic bleakness of
Tierra del Fuego. In between lay every kind of climatic extreme you
could think of—the rain forests of Brazil, the deserts of New Mexico,
the mountains of the Andes, and much else. All succumbed to these
hominid arrivistes, but when? Well, some time between 10,000 and
30,000 years ago is the best that archeologists can come up with.

The first clear evidence of human occupation in the Americas
comes in the form of Clovis spear points, the earliest of which have
been reliably dated as being about 12,000 years old.[18] These stone
tools have been found across the United States (but not in Canada,
which was mostly smothered in glaciers at the time) and are named
after the town of Clovis, New Mexico, near the Texas border, where
they were first uncovered. The Clovis people were probably some of
the finest human hunters thrown up during mankind's evolution and
appear to have been in incessant movement. They camped along
rivers, beside streams, close to waterholes, and hunted elephant-like
mammoths and mastodons, bison, horses, and enormous giant
ground sloths, in competition with lions, giant wolves, and saber-
toothed cats. They butchered their prey where it fell, and used lightweight tools made of fine stone points that are described as being
fluted because they have a groove running their lengths. This channel
was carved either to help bind the implement to a spear, which
would have been thrown by hand, or to a shaft, which would have
been propelled by a throwing stick or a bow. Whichever was the case,
it proved to be an extremely effective technology, lightweight or
not—for mammoth and bison skeletons have been uncovered with
Clovis spear points buried deep inside their rib cages. In one case, a
skeleton from southern Arizona was found with a total of eight such
blades embedded in it. The Clovis people were mighty hunters and
highly effective colonizers, as can be judged from the fact that by
11,000 years ago, humans had spread to both coasts of America, and
from the area we now call the Midwest to the tip of Patagonia in
South America.

Now this expertise and resilience raises a very important issue, for

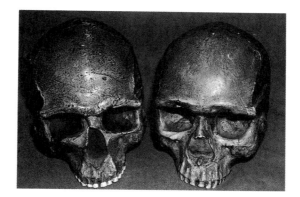

42 This Upper Cave skull from Zhoukoudian (right) is one of the oldest known modern human fossils from China. Statistical tests show it does not look "Chinese," but resembles skulls from Australia, Africa, and Europe, such as the Cro-Magnon skull from Predmostí, Czech Republic (left).

the emergence of the Clovis people, with their exquisitely chiseled spears of translucent chert, coincides almost exactly with one of America's major bouts of extinctions. For many years this calamitous drop in species—nearly all of which were large mammals—had puzzled scientists. They knew it had occurred. They had, after all, uncovered the remains of these great creatures that once strode the plains of middle America. "It is impossible to reflect on the state of the American continent without astonishment," noted Darwin. "Formerly it must have swarmed with great monsters; now we find mere pygmies."[19] The trouble was no one was sure when this great eradication had taken place or how long it had lasted.

Then, in the 1960s, came the archeologists with their latest weapon: radiocarbon dating. They looked at the bones of those fallen mammoths, extinct mastodons, and all the rest, and found they had disappeared with extraordinary rapidity, in some cases in less than 300 years, and all roughly around 11,000 years ago—just as the Clovis folk began to sweep down North America. "Large animals disappeared not because they lost their food supply, but because they became one," said one of the leading investigators, Paul Martin of the University of Arizona.[20]

Martin became (and still is) convinced that the Clovis people were responsible for the fact that at least seventy-five species—including woolly mammoths, mastodons, four-horned antelopes, llama-like mammals, capybaras the size of Newfoundland dogs, and lumbering sloths the size of giraffes—vanished off the face of the continent in an astonishingly brief time. And in the wake of these herbivores followed the animals that had preyed upon them: a species of North American lion, for example, and the saber-toothed tiger.

In South America, this slaughter was repeated, with creatures like the glyptodon (a giant armadillo-like animal), several species of large rodent, various llama and pigs, as well as many of those mammals that had already been wiped out in North America. A similar bloodbath occurred in Australia when many large members of its unique fauna were expunged from the environment. This time, however, the extinctions occurred much earlier, around 30,000 years ago, after humans had already spread across the continent. In the Old World, by contrast, no such bout of mammal massacring could be detected. Large creatures there had either perished long ago at the hands of *Homo sapiens'* predecessors or had learned from experience to avoid them, it was argued. (The statistics are as follows: over the past 100,000 years, North America has lost 73 percent of its large mammals; South America, 79 percent; Australia, 86 percent; but Africa, only 14 percent.)[21]

Further support for Martin's overkill hypothesis, as he called it, was put forward by the Scandinavian paleontologist, Bjorn Kurtén. He noted that although not all North American large animals had died at the time of the Clovis people, most of those that had survived shared one characteristic—they had arrived late on the continent, over the same Beringian land bridge that humans had crossed. These Asian fellow-travelers had long experience of men and women, and the deadly weapons they used. "It is noteworthy that most of the Eurasian invaders of North America—the moose, wapiti, caribou, musk ox, grizzly bears and so on—were able to maintain themselves, perhaps because of their long previous conditioning to Man," said Kurtén.[22]

Not every scientist agrees with this notion of a blitzkrieg of human hunters, equipped with lances, arrows, and spears, overwhelming these previously uninhabited lands, eradicating every major form of game they could see. For example, Don Grayson of the University of Washington points out that just because we cannot find large mammal fossils in archeological sites that are younger than those of the Clovis era, this does not mean they do not exist and that mammoths or mastodons did not thrive on for many more millennia.[23] In addition, it is an assumption—and no more than that—that the killer of one species was responsible for wiping out the rest. Climatic changes, triggered by the passing of the Ice Age, which brought warmer but more extreme weather, were the real culprit, he says. Other skeptical scientists simply cannot imagine how these primitive people, armed with only stone spears, could kill creatures as big as a North American mammoth.

But this was not a business of just hurling a few boulders or spears at giant creatures and then running off if the plan failed, Jared Diamond points out. Modern Africans and Asians, often hunting alone and using only a spear or poisoned arrow, stalk and occasionally kill elephants. "These modern elephant hunters still rate as amateur dabblers, compared to the mammoth hunters of Clovis times, heirs to hundreds of thousands of years of hunting experience with stone tools," adds Diamond. "Instead a more realistic picture is of warmly-clad professionals, safely spearing a terrified mammoth ambushed in a narrow stream bed."[24]

As to the idea that climate was the cause of all those extinctions, this is given short shrift by Peter Ward of the University of Washington, in his authoritative study on the issue, *The End of Evolution*:

There is no doubt that the end of the Ice Age was accompanied by sudden and drastic changes in temperature, and that a dramatic change in plant communities and their distribution across the North American continent occurred soon after. But the idea that all the larger mammals were unable to migrate out of harm's way seems unlikely; we know that many large African mammals are perfectly capable of making long treks in search of seasonal food sources or water. Climate change alone seems unlikely to have killed off 35 genera of North American mammals so rapidly.[25]

Certainly, given our track record in dealing with the animal world at other times (see Chapter 9), it would seem fair that humans should shoulder most of the blame for devastation of the fauna of North and South America.

The Clovis people were presumably the ancestors of some present-day Native Americans, though it is unlikely the continent's colonization was a single event—archeology, the diversity of present-day native languages, and patterns in teeth, all point to the fact that several waves of immigration must have taken place. Nor were the Clovis folk necessarily the first Americans. People of earlier migrations may well have been replaced by them in some regions, but still gave rise to descendants elsewhere—people such as the Ona, whom Charles Darwin encountered in Tierra del Fuego and whom he thought singularly debased. "Their red skins filthy and greasy, their hair entangled, their voices discordant, their gesticulation violent and without any dignity. Viewing such men, one can hardly make oneself believe that they are fellow creatures placed in the same world," he wrote.[26] Nevertheless, they were members of *Homo sapiens* (as Darwin readily admitted when in a less bilious mood), though they looked so different from other Native Americans that some anthropologists suspect they were the last descendants of an earlier migration. Purebred Ona are, sadly, extinct and the idea can only now be tested from DNA left in their skeletons.

In fact, it is genetic evidence that provides the greatest challenge in our attempts to pinpoint the date of our first steps on to a continent that now so dominates life on earth. While the Clovis tools and other remains indicate dates around only 15,000 years ago for humanity's entry into the Americas, some archeologists suggest we may have arrived there as long ago as 35,000 years, an idea given support from genetic data gathered by Cavalli-Sforza and his team.[27] His analyses of Native American blood and proteins indicate these are divergent enough to suggest the continent was settled thirty millennia ago, and in at least three different waves of immigration.

This idea is supported by studies carried out by Doug Wallace, of Emory University.[28] He and his colleagues have found that one of four rare variants of mitochondrial DNA in Native Americans is also

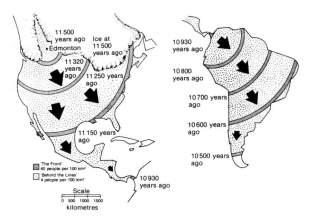

43 Martin's view of the spread of humans through the Americas, and the associated waves of mammal extinctions.

found in Asians, but not Europeans or Africans, clearly indicating their origins. More to the point, the frequency of these variants is much higher in Americans than in Asians, suggesting the former are descended from a smaller number of "founding mothers" from Asia. This number could be as small as four women present in the first group, or it could simply be that there were four groups of closely related women in this hunter-gatherer vanguard. Either way, there was a clear bottleneck in population occurring at this time—which the group dates as being between 42,000 and 21,000 years ago.

Some archeological support for a longer occupation of the Americans has been found at South American sites, such as Pedra Furada in the dry thorn forest country of northern Brazil. This site, beneath a high sandstone cliff, has revealed stone tools and hearths dated by radiocarbon with decreasing reliability from less than 10,000 to nearly 50,000 years ago. The latter date suggests a pedigree of some distinction for America. However, experts have challenged the claim that the hearths are man made and have argued that the oldest implements are really just cobbles that had fallen from the cliffs to smash at the bottom, mimicking the hand of an ancient, but in fact nonexistent, American settler.[29]

So was America first colonized by a few waves of wanderers who left Asia 15,000 years ago, or are their origins at least twice that age? It is still a baffling question. Nor do the anthropological headaches get any easier when we turn to Australia, one of the most mysterious of all human homelands.

Australia is a vast stretch of land, still largely devoid of people. It has tropical forests in the north, an arid heartland, and cool woods. Until recently, it even had ice sheets in the south. In the past, it was filled with rich hunting territories that sported many species of large birds and strange marsupials: ten-foot-tall kangaroos; the diprotodon, a sort of browsing, rhinoceros-sized wombat; a lion-like marsupial carnivore; giant koala bears; deer-like marsupials; and a giant monitor lizard, the size of a horse. This unique fauna had started evolving in isolation from the rest of the world when Australia separated from South America and Antarctica more than forty-five million years ago. Then *Homo sapiens* appeared on the scene, our arrival coinciding more or less with the abrupt extinction of all this exotic fauna—though through which stage door, and at what moment, humankind chose to make its dramatic entrance, we can only guess.[30] Certainly, their route could not have been an easy one. The islands of southeast Asia were mainly covered with thick jungles and even at the lowest sea levels of the Ice Age, with Tasmania, Australia, and New Guinea lumped together in a single continent, Asia was a considerable distance away. Men and women would have had to navigate many different journeys between islands separated—in some cases—by forty miles or more of open sea. So surely only a sophisticated, cultured, and therefore recent society made this epic crossing, archeologists assumed. This temporal chauvinism was to receive a nasty jolt, however.

The Willandra region of south Australia is today just shrub and desert, though once its lakes teemed with fish and shellfish, and giant marsupials wandered round their shores, as did early Australians. In the last case, we know of their presence because several human skeletons were discovered in 1968 near the dried-out bed of Lake Mungo. One was a corpse that appeared to have been partially cremated. The bones were burned and broken, and as the individual had been less than five feet tall and lightly built when alive, it was accepted this was the skeleton of a woman. Another body, deemed to be male, had been buried stretched out and coated with red ocher, like many Cro-Magnon burials. Radiocarbon dating was applied and revealed the cremation was about 26,000 years old—the oldest example of this practice yet discovered. As to the burial, it was found to be over 30,000 years old.[31]

These findings were surprising. For one thing, they indicated that seafaring—the only means by which the settlers of Australia could have arrived—must be a skill that goes back much further into the prehistory of our species than was previously supposed. The people who reached Australia had built seafaring boats at least 20,000 years before such craft appeared elsewhere in the archeological record. In addition, the red ocher, which had been carried from several mines to the site, suggested pigments were being used for body decoration and painting around the same time as Cro-Magnons are known to have carried out this practice. The remains of fish, shellfish, crayfish, egg shells, small birds, and mammals, as well as stone tools, scattered around the sites, also testify to a sophisticated Stone Age lifestyle. Archeologists even found ovens—pits in the sand full of ash and charcoal, capped by baked clay. This was no evolutionary backwater.

Scientists were still absorbing the Mungo data when archeologists and paleontologists stumbled on an even stranger set of human remains at Kow Swamp several hundred miles to the south. A number of partially preserved skeletons were unearthed, some again showing clear evidence of cremation and the use of red ocher. But this time, the burials were found to be more recent—about 10,000 years old. Interestingly, some of its former inhabitants looked very different from those at Mungo, displaying big faces, jaws, and teeth, strong browridges, and flat foreheads. The discovery was startling to say the least, for the delicate people of Mungo seemed to precede the more robust, heavier Kow Swamp inhabitants, a reversal of the usual progress of recent human evolution.

So what was going on in Australia 10,000 to 30,000 years ago? Well, according to multiregionalist Alan Thorne, the continent must have possessed two distinct populations of humans. One (the Mungo people) was gracile, and delicately built even compared with present-day Australian Aborigines. These people migrated from China, carrying the genes of Peking Man via New Guinea and eastern Australia,

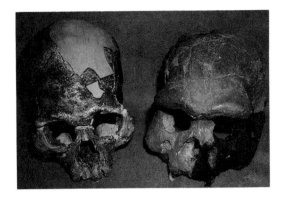

44   This *Homo erectus* skull from Java (right) is regarded by multiregionalists as
     possibly ancestral to the Australian skull from Kow Swamp (left). In an "Out of
     Africa" scenario, the skulls instead represent entirely separate radiations of
     people from Africa, at least a million years apart.

he argued. The others, the heavily built Kow Swamp inhabitants,
arrived via Java, following a western island route to Australia (down
through Sumatra, Timor, over the sea to Australia's Northern Territory, before sweeping round its west coast), carrying the genes of the
Java and Ngandong people with them. Both these early colonizers
must have remained separate for at least 20,000 years, before
merging about 10,000 years ago to become the ancestors of present-
day Aborigines. In other words, the eventual blending of smaller,
gracile Mungo people, with larger, robust Kow Swamp folk, produced moderately-sized modern Aborigines.[32]

Not everyone agreed with this explanation, however. Peter
Brown, a former student of Alan Thorne, working at Australia's University of New England, questioned the basic assumption that the
two, critical gracile Mungo fossils represent a male and a female.
Yes, the cremated body is a female's, he says. But no, the other,
buried corpse is not a man's. It too is female. Now this may seem a
mere paleontological quibble, a matter of academic dispute over a
fossil interpretation. Not so, for if both skeletons are those of
women, then this implies that the Mungo folk only look gracile
because our fossil sample is made up of lightly built females. Mungo
men could easily have looked heavier, like the Kow Swamp people,
implying that Australia did not support two Stone Age populations

of gracile and robust people. There was only "a single, homogenous, Pleistocene population" as Brown put it. And those flat foreheaded skulls found at Kow Swamp? These, he argued, were artificially deformed by head binding or by using a strap running across the forehead to carry loads on backs—practices known to have affected skull shapes elsewhere.[33]

As to the thickening of the skull walls which is supposedly present only in the robust Australian group, Brown has come up with a decidedly exotic explanation, based on observations of contemporary Aboriginal customs. A traditional way to settle land or property disputes is to use heavy wooden clubs. Aggrieved parties face each other, in turn striking or attempting to parry blows. The disagreement is finally settled when one protagonist is wounded seriously enough to be disabled. As such conflicts are most common between young adults of prime reproductive age, individuals who inherit thick skulls would be favored, says Brown. Remarkably, about half of all Aboriginal skulls from southern Australia show deep wounds either on the front or on the side of the head (most often on the left side, matching those that would have been inflicted by a right-handed opponent). There is a similar level of healed forearm fractures, the consequences of parrying blows. This pattern is present even in 11,000-year-old skeletons which show head and forearm fractures, Brown has found. Far from being primitive *Homo erectus* features, these thick skulls are therefore the consequence of the ritual clubbing that remains the Aboriginal equivalent of the small claims court, he argues.[34]

This general interpretation, of a single population of ancestors slowly evolving into Aboriginal people today, is now favored by many anthropologists such as Colin Pardoe, of the South Australian Museum, Adelaide. As he puts it:

A model of diversification must be seen to be the more appealing than a model of multiple origins on the ground of parsimony, the broadest explanation of data and evolutionary theory, especially that concerning gene flow. The conflation of sexual dimorphism and robusticity is readily apparent, while the complexity of a migrationist approach necessary to separate two

45   The Tasmanian William Lanne.

founding populations for thousands of generations cannot be countenanced.[35]

In keeping with the best tradition of modern paleontology, the dispute between the Brown and Thorne camps has become increasingly heated and bitter in recent years, with the feud coming to a head over a highly controversial study of a 14,000-year-old skeleton found in a cave on King Island which lies between Tasmania and Australia. This is an important site that should provide evidence of links between the island and mainland Aborigines before sea levels rose 10,000 years ago. But strict laws covering the examination of possible ancestral Aboriginal burials have been imposed recently and Thorne and his colleague Robin Sim were allowed only three hours to study the skeleton before it was reburied. They managed to take thirty measurements, and concluded it was a member of the gracile Mungo people, in keeping with Thorne's two-population model. Such an interpretation was an anathema to Brown, however, who published a vociferous attack on the work on the grounds that the skeleton's reburial meant no one else could check the accuracy of their observations. He argued that most of Thorne's observations actually indicate that the King Island skeleton is a woman's, backing his own theory. Obviously, the solution would be to study the skeleton again. However, permission is unlikely to be given, for the whole issue has become even more confused since the introduction of further reburial laws in Australia, a move which also led to the reinterment of the entire Kow Swamp collection, on which Alan Thorne based his ideas of a "robust" prehistoric population.[36] It is a highly inflammatory topic—and a vexing one—for anthropologists and paleontologists.

And, of course, one sympathizes with this plight. It is infuriating to know of the existence of evidence that could help prove or disprove a scientific problem. On the other hand, the evils perpetrated by Europeans upon the Aborigines, particularly those from Tasmania, makes it impossible not to understand their sensitivity over this issue. When discovered in 1642, Tasmania supported about 5,000 hunter-gatherers. They made simple stone and wooden tools, but apparently lacked—unlike their mainland cousins—boomerangs and nets. Then the white settlers came and began kidnapping children for laborers, and women for consorts. The men were simply killed. In 1828, martial law was declared, and soldiers were ordered to shoot on sight any Aborigine in a settled area. Two years later, the last purebred Tasmanians were rounded up, and transported to nearby Flinders Island. The site was run like a jail. Poorly fed, most inmates died, and few infants survived more than a couple of months after birth. The last man, William Lanne, died in 1869. The insults persisted, however. Scientists fought over his body, which they claimed was a missing link between apes and humans. His corpse was continually dug up and reburied, parts being removed each time—head, feet, hands, ears, nose, etc. One doctor even made a tobacco pouch out of Lanne's skin. The last woman, Truganini, who died in 1876, terrified that she might suffer similar mutilations, asked to be buried at sea. Her plea was in vain. Her skeleton was exhumed and displayed in the Tasmanian Museum. Finally, in 1976, one hundred years after her death, her bones were cremated and buried at sea as she had requested. Given such grotesque indignities, science should not be too surprised that natives of many countries—Tasmania and Australia, Hawaii, and mainland America, for example—are now trying to reclaim their history in a manner that is antagonistic and unsympathetic towards modern scientists who are trying to study their origins.[37]

In the meantime, researchers have continued to produce their paleontological surprises. Thermoluminescence studies of grains of sands from deposits at the North Australian sites of Malakunanja II and Nauwalabila, which have many stone tools and crayons of red ocher embedded within them, suggest they are 50,000 to 60,000 years old—the earliest known in the continent. This research also indicates settlers arrived from the west, and that they may already have been moving into the arid zones of the continent a very long time ago.[38]

If nothing else, all these recent studies show that modern humanity's outward urge was a very old one. But why did *Homo sapiens* come to places like American and Australia in the first place? Can we envisage them peering across the snowy landscapes of Beringia or the open seas of southeast Asia, and wondering what lay beyond? Most probably they did not. The pressure to move was more likely to have been motivated by population growth. New generations needed new foraging territories because the land simply could not support high densities of hunter-gatherers. As Kingdon points out:

> The movement or expansion of people over considerable distances is often imagined in individualistic terms, as if prehistoric groups were seized by the urge to explore or migrate. Such movements did not depend on individual wills; it was external events that imposed constant change and flux on human existence. A succession of bad years, incursions by aggressive neighbours, overpopulation, overhunting, the invention of a new and superior technique, fleeing from disease or fulfilling the prophecies of a shaman; all these and more could have triggered movement on and into the unknown.[39]

The first people reaching the Americas were therefore unaware of the momentous voyage they had made. They would most likely have been following migrating herds of reindeer across Beringia. On the other hand, the first humans to reach New Guinea or Australia would have understood very rapidly that they were somewhere terrifyingly new, and that they probably could never return to their homeland. While the first Americans would have seen familiar plants and animals in Alaska and Canada, the first Australians really did arrive in a New World filled with strange creatures. Nor should we assume their journey to the continent was a simple one of island hopping at times of low sea levels. They may have taken place when seas were high. Rising waters would have shrunk habitats, increasing population pressures. To escape, groups probably set off for land they could see, but found themselves swept away by unforgiving shifts in tides and winds. Many of these ancient boat people perished. Nevertheless, some survived, to find themselves washed up on the shores of a strange land—on which they founded a whole new race of people.

It is a story that has been repeated countless times, as the world has been convulsed with continual waves of migrations and invasions, eventually producing a planet peopled by white-skinned, blue-eyed Scandinavians; enigmatic Basques, with their strange language and their distinctive blood patterns; the Baika Pygmies of Central Africa; the teeming tribes of New Guinea; Samoans; the Falasha Jews of Ethiopia; the Amazon's Yanomamo people; the Tiwis of Australia; and hundreds of others. In the past, some scientists, philosophers, and historians have made much of the intrinsic differences between these groups, linking them with all sorts of stereotypes—meanness, efficiency, laziness, and others. But it is a quite specific corollary of the Out of Africa theory that such ideas are outdated. The progeny of the people who found Australia 50,000 years ago, and the descendants of the tribes who poured down the Americas 12,000 years ago, as well as the heirs to all those other settlers of Europe, Africa, and Asia, share a common biological bond. They are all the children of those Africans who emerged from their homeland only a few ticks ago on our evolutionary clock. They may have diverged geographically since then, and developed superficial variations, but underneath our species has scarcely differentiated at all. We may look exotic or odd to our neighbors in other countries, but we are all startlingly similar when judged by our genes. Yet the issue of racial differences continues to dominate world affairs. Serbs fight Bosnians, Tutsi slaughter their Burundi neighbors, and blacks and whites keep an uneasy peace in downtown America. This divisive schism has been the source of untold misery for thousands of years. Yet our new evolutionary perspective offers us an opportunity to reexamine its roots and its implications—as we shall see.

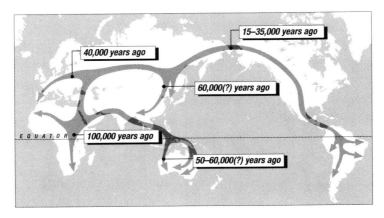

46   Genes and fossils have been used to reconstruct this map of the spread of *Homo sapiens* over the last 100,000 years.

7

Africans Under the Skin

We ignore the intimate relatedness of all humans when we look for and amplify the minutest distinctions among us, until we find ourselves surrounded by what looks more like our natural enemies than members of our own close-knit species.

Erich Harth, Dawn of a Millennium

How far back is our childhood? I think our childhood goes back thousands of years, farther back than the memory of any race.

Ben Okri

Sir Philip Mitchell, former Governor of Kenya, had a low opinion of the African nations that formed the backbone of the British Empire. These countries were populated by "people who had never invented or adopted an alphabet or even any form of hieroglyphic writing," he wrote in the 1950s:

They had no numerals, no almanac or calendar, no notation of time or measurements of length, capacity, or weight, no currency, no external trade except slaves and ivory . . . no plough, no wheel and no means of transportation except human head porterage on land and dugout canoes on rivers and lakes. These people had built nothing, nothing of any kind, in any material more durable than mud, poles, and thatch. Great numbers wore no clothes at all; others wore bark cloth or hides and skins.[1]

It is a distinctive diatribe: a process of denying the achievements of other peoples, in this case the inhabitants of Europe's colonies, and of exaggerating their shortcomings in order to show they were too backward to manage their own affairs. It is an attitude that echoes the words of Thomas Hobbes writing in *Leviathan* in 1651. "No arts; no letters; no society and which is worst of all, continual fear and danger of violent death; and the life of man, solitary, poor, nasty, brutish and short."

It is not an outlook shared by those who have taken the trouble to study the lives of such individuals, however—researchers like anthropologist Germaine Dieterlen. Compare her views with Sir Philip's:

The Africans with whom we have worked in the region of the Upper Niger have systems of signs which run into thousands, their own systems of astronomy and calendrical measurements, methods of calculation and extensive anatomical and physiological knowledge, as well as a systematic pharmacopoeia. The principles underlying their social organization find expression in classifications which embrace many manifestations of nature . . . plants, insects, textiles, games and rites are distributed in categories that can be further divided, numerically expressed and related one to another. It is on these same principles that the political and religious authority of chiefs, the family system and juridical rights, reflected notably in kinship and marriages, have been established. Indeed, all the activities of the daily lives of individuals are ultimately based on them.[2]

These two visions of African life contrast with each other vividly, the latter, we would hope, being the one now that holds most sway—though antediluvian views, like Sir Philip's, still influence modern attitudes. Even today, atrocities in countries like Uganda or Rwanda are sometimes reported with an unspoken commentary: "What can you expect? Africa will never be able to govern itself in a 'civilized' fashion." Yet only five decades ago, one of Europe's most "civilized" nations systematically exterminated six million people and dragged the world into a war that killed at least another forty million. Nor

47   Andrea Searcy (far right) is the daughter of an African-American and a Native
     American. With skillful makeup and contact lenses she can readily look
     African, European, or Oriental. Selection could certainly have duplicated such
     a process over the last 100,000 years to produce today's racial differences.

have conflicts in Bosnia and Serbia helped to portray Europe today as
an entirely "civilized" continent. Africa is no worse, and no better,
than any other global arena, for the simple reason that human cruelty
is universal and knows no geographical boundaries.

Stereotyping according to races runs deep, however, even in science, having once been thought to represent deep biological divisions between the peoples of the world. For decades, *Homo sapiens'*
global divergences were assumed to be the vestiges of million-year-
old cleavages in our family tree. Race has a profound biological
meaning, it was reckoned. Recent acceptance of the Out of Africa
theory has changed that perspective—for it has been shown we are
indeed all Africans under our skin, and that our differentiation into
Eskimos, Bushmen, Australians, Scandinavians, and other populations has merely been a coda to the long song of human evolution.

Just consider the data. There were the discoveries, at Katanda in
Zaire, made by John Yellen and Alison Brooks: bone harpoons, delicately carved knives, the evidence of systematic fishing, and the clusters of stones and debris that suggest huts were being built 90,000
years ago (see Chapter 1).[3] All these were created thousands of years
before *Homo sapiens* got round to similar work in Europe. In other
words, Africa was no cultural cul-de-sac, despite the views of Sir
Philip and his ilk.

Then there is the genetic evidence. It reveals the stark, simple,
homogeneous nature of modern mankind. Scientists have generally

48   Arnold Schwarzenegger modified by a computer to look African-American.

recognized that the common chimpanzee of central Africa has three subspecies, though to most people they look very similar indeed. Nevertheless, these chimp "races" are almost ten times as different from each other, genetically, as are the African, European, and Asian divisions of *Homo sapiens* created by Linnaeus, Blumenbach, and Coon (see Chapter 3).[4]

In addition, there is the work of biologist Richard Lewontin of Harvard University, who used a series of seventeen stretches of DNA to study the differences between 168 populations, people such as the Austrians, Thais, and Apaches. In this way, Lewontin found there is more genetic variation within one race than there is between that race and another.[5] Only 6.3 percent of the dissimilarity between two people from disparate ethnic backgrounds can be explained by their belonging to separate races. This means that if you pick at random any two people, say Norwegians, walking along the street and analyze their twenty-three chromosomes, you could find their genes have less in common than do the genes of one of them with that of a person from another continent. As Sharon Begley puts it in an essay on race in *Time* magazine in February 1993: "Genetic variation from one individual to another of the same race swamps the average differences between racial groupings." Categorizing a person by their race can clearly be a deeply misleading business.

Genes do reveal some differences between populations, of course. For example, mitochondrial DNA studies have shown that human diversity within Africa is nearly three times that in Europe, and nearly double that within Asia.[6] This same high level of African differentiation is apparent in studies of skull measurements,[7] and in a number of recent nuclear DNA studies such as those of Kidd and Tishkoff (see Chapter 5). Whether Africa's greater variation of populations is a reflection of its deeper antiquity, or of its earlier recovery in numbers from the bottleneck which preceded the global spread of modern humans, remains unclear.

Nevertheless, the message from the Out of Africa theory is a straightforward one. Our exodus's timescale is so brief that only slight differences, if any, in intellect and innate behavior are likely to have evolved between modern human populations. And yet there are those who would still deny this fact, workers who cling to an intellectual tradition that runs from Galton to Eysenck and Jensen, scientists who have argued that racial differences in psychological and intellectual abilities are deep and meaningful. Indeed, the subject has been fueled to near ignition in recent years thanks to claims that researchers have uncovered new evidence that the world's populations are easily separable into distinct categories, particularly with regard to intelligence.

Take the example of British-born Philippe Rushton of the University of Western Ontario. In a paper presented at the American Association for the Advancement of Science in 1989,[8] he used the Out of Africa theory as a starting point for arguing that humans evolved in a rich but unpredictable environment (Africa) where natural selection favored a strategy of high birthrates and low levels of parental care. However, some populations moved into more challenging environments (Europe and Asia) where lower birthrates and more intense parental care were favored. Whites and, particularly, Orientals, have evolved this adaptation, says Rushton, while black people retain our ancestral African pattern. He ranks these races for features such as brain size, educational and occupational achievement, rate of maturity, fecundity, promiscuity, penis size, aggressiveness, parental care, respect for the law, and many others, revealing blacks to be the most primitive, and Orientals the most advanced, with whites lying somewhere in between. For example, Rushton reports that IQ tests consistently show a grading in favor of Orientals (average score 107), followed by whites (100), with blacks some way behind (85).

The entire thesis is based on some very odd suppositions, however, For one thing, Rushton simply assumes that the Out of Africa theory establishes the primitiveness of Africans and the superiority of Orientals. It does nothing of the sort, of course. As we have seen throughout this book, indeed this chapter, the theory provides no rationale for supposing that Orientals are evolutionarily superior beings or blacks inferior. It would be expected that Europeans and Orientals are more closely related through recency of common ancestry, but there is no obvious basis for Orientals being more "evolved" than Europeans or blacks. And while we know that the supposed ancestors of Europeans—the Cro-Magnons—lived through the peak of the last Ice Age, we have little data to show where Orientals evolved. Equally, the relative merits and drawbacks of African and non-African habitats are certainly arguable. Is surviving equatorial drought more or less challenging than enduring an Arctic winter, for example?

Then there is the issue of Rushton's data and disparate sources. One of the most heavily cited in his earlier works is an 1898 book, by an anonymous French army surgeon, a repository of anecdotal information about the penis, breast, and buttock size of different native populations. His reliance on these kinds of data led to some strong criticism. More recently, Rushton has used a wider variety of sources, including census data, studies of military recruits, and brain-scanning techniques.[9] He argues that such data reinforce his original conclusions that human populations can be ordered in a scale of evolutionary advance from blacks, to whites, to Orientals.

Not surprisingly, Rushton's work has provoked a storm of protest.

He has been dubbed a racist and has been threatened with removal from his teaching post. However, his university has so far maintained his right to academic freedom of expression. More intriguingly, Rushton has also received more than $700,000 from the Pioneer Fund, set up in 1937 to further "race betterment, with special reference to the people of the United States." In fact, the fund, established in 1937 by Wycliffe Draper, a Harvard graduate who inherited a textile machinery fortune, has had a highly controversial history, having backed research programs on human behavior and variation at more than sixty institutions in eight countries. These projects include several that have claimed to have uncovered strong links between race and intelligence.[10]

Rushton's views might have remained on the fringe had they not received a sudden, invigorating dose of intellectual oxygen with the appearance of *The Bell Curve*, by the late Richard Herrnstein and Charles Murray[11]—a book whose publication in 1994 triggered widespread controversy in the United States. (The bell curve of the title refers to the characteristic shape of a diagram of a trait, such as IQ, in a population in which few people display exceptionally low or high scores while the majority cluster near the average value. Such a dispersion—known as a normal distribution—produces a graph that looks like a bell.) Herrnstein and Murray, who cite Rushton's work eleven times, as well as that of many other Pioneer Fund beneficiaries, concentrate their arguments on the over-representation of American blacks in the country's lowest social levels and as scorers of its lowest IQ ratings. They conclude that no amount of state or national aid could achieve real racial equality in intellect and achievement. The book would have been incendiary at any time and in any country. However, its appearance during a period when Americans, including many liberals, had begun to question the merits of the nation's welfare system and to challenge its civil rights agenda, hit an exposed nerve. Indeed, such was the fuss that when the *New Republic* magazine prepared to print an eleven page extract from the book, an army of columnists and staff writers protested so strongly that a total of sixteen pages of rebuttals also had to be published.[12]

Much of this sort of work is based on the idea that there is an indisputable correlation between IQ scores and brain size. Rushton, for example, makes a great deal of the fact that a woman's brain size is smaller on average than a man's, as is a black person's compared to a white's or an Oriental's. According to him the issue is straightforward, though it is in reality extremely complex: because brain size statistics are variously based on estimates of external head dimensions, external skull dimensions, internal skull dimensions, and volume and weight measurements on the actual brains of people who have died. In addition, the relationship of brain size to body size has to be taken into account since there is a clear correlation (as there is with all primates) between the two. Quite simply, bigger bodies need larger brains to govern them. For example, a woman's brain is on average 13 percent smaller than a man's brain. But when scaled against her smaller body size, the difference disappears. (Rushton does not accept this argument, however, and introduces a new factor—fat—into his equations. He maintains that women have about 20 percent fat in their bodies and men only 10 percent, he uses these "fat allowances" to show women really do have smaller brains than men.[13] Not surprisingly, this argument has not gone down well in many circles.) If nothing else, this problem shows the difficulties of trying to study body and brain size relationships. In the former case, should we use lean body weight, total weight, height, or surface area? You take your choice . . .

As to racial differences, it is well-known that there is significant global variation in brain size. For example Tierra del Fuegian men have an average cranial capacity of 1,590 milliliters, while a sample of Peruvian women gives 1,219. Similarly, French men produce a figure of 1,585; women from the Tyrol, 1,238; men from the Xhosa (Nelson Mandela's tribe), 1,570; and Kenyan women, 1,207.[14] So why these variations? Well, as we have seen, people living near the tropics are lighter bodied, on average, than those who dwell near the poles. This is a specific case of Bergmann's rule which states that animals tend to be larger (in particular, rounder) in colder climates to help to conserve body heat. Equally, people in warmer conditions have smaller (mainly thinner) bodies. And, of course, because brain and body size are linked, they will also have smaller brains. Put simply: hot weather means smaller but longer brains, while a cold climate produces larger, rounder ones. And that indeed is what we see. The largest study of global cranial volumes ever carried out, by Beals, Smith, and Dodd in 1984, showed that climate of origin was the most important variable influencing the size of the human cranium. "Any effort to contribute racial or cognitive significance to brain size is probably meaningless unless climate is controlled," they state. "For example, the endocranial [inside the braincase] volumes of Europeans and Africans differ little from what one would expect given the difference in their respective winters."[15]

For their part, Rushton and the like accept that people living in colder climates have bigger brains, while those from the tropics have smaller ones, but for reasons that go beyond the issue of body size. They argue that brains grew to cope with the harsher, more challenging conditions involved in living in the higher latitudes. "The further north the populations migrated out of Africa, the more they encountered the cognitively demanding problems of gathering and storing food, gaining shelter, making clothes, and raising children successfully during prolonged winters," states Rushton.[16] "As the original African populations evolved into Caucasoids and Mongoloids, they did so in the direction of larger brains and lower levels of sex hormone, with concomitant reduction in aggression and sexual potency and increases in forward planning and family stability." In other words, our brains got bigger because we needed more intellect to deal with life in Europe, Asia, and Australia, and the Americas later on.

But there is a crucial flaw in this reasoning. It assumes that intellect and brain size are intimately related, and that is by no means clear. Just consider the case of two of the cold-adapted populations that we encountered in the previous chapter—the Ona, the original settlers of Tierra del Fuego; and the Tasmanians. They are useful benchmarks for our comparisons of brain size and achievement because Fuegian skulls and brain capacities are among the largest of all *Homo sapiens* while the average Tasmanian brain size was higher than that of their mainland Australian cousins, as we might expect.

Consider Tierra del Fuego first. Its inhabitants, the Ona, lived in land that lies just outside the Antarctic circle, at the very southern tip of South America. Surrounded by the Atlantic, Pacific, and Antarctic oceans, the archipelago is swept by rain and blizzards: ripe, challenging territory for growing those large skulls so admired by Rushton and the rest. And this cranial growth is indeed what took place, with most scientists maintaining that it was related to an increase in body size. For Rushton et al., this enlargement must also reflect increased intelligence, however.

But if the latter scenario is correct, how can we reconcile the advanced cerebral status of the Ona with their impoverished existences. Readers will recall Darwin's description of their low, wretched lives, unadorned by fire or decent clothing. "Viewing such men, one can hardly make oneself believe that they are fellow creatures and inhabitants of the same world," he wrote.[17] Yet these folk, according to our recent racial demographers, should represent the very acme of human achievement, endowed with large brains and living—as they undoubtedly did—in one of the most challenging of cold climates.

Then there were the Tasmanians. Archeological evidence shows that 20,000 years ago they were making piercing tools from wallaby bones, and necklaces and engravings. Then their Tasmanian homeland became isolated by sea level rises 10,000 years ago. Slowly their tool kit became simpler and simpler, while their smaller-brained mainland cousins, living in an apparently "debasing" warmer climate (according to Rushton that is), produced—about 6,000 years ago—a sudden leap in implement technology that was one of the great flowerings of Stone Age culture.

In short, having larger crania and bigger brains within them, did little for the Fuegians' or Tasmanians' quality of life. In fact, they were neither more nor less intelligent than the rest of humanity, as Darwin discovered when he met Fuegians who had been "civilized"—i.e., acclimatized—by spending a year in England. By this time, they spoke good English, dressed in Western clothes, and were considered to be sophisticated enough to meet the Royal family. What impoverished their lifestyles in their homeland was cultural isolation, not lack of brainpower. Like the Tasmanians, the Fuegian people were stuck at the lonely, southern end of a large continent. And this is the critical point quite overlooked by the new racial evangelists. They assume large brains mean big intellects, and the reverse. Yet, it is not that simple. We would not expect a person to be clever because they wore a large hat, after all. Moving from general statistics to judging individuals is always dangerous. Other workers—such as Majie Henneberg of Adelaide University—suggest that human brain size is in fact a poor predictor of achievement. His work has revealed a low correlation of IQ with brain size,[18] and more to the point suggests that if the opposite were true—i.e., that brain size and intellect were in fact linked—our species must be getting stupider by the millennium. It sounds extraordinary. Nevertheless, Henneberg and others have discovered that human brain size has decreased almost universally over the past 10,000 years, an absorbing story whose telling forms an important part of the final chapter of this book, and for which hungry readers will have to wait (or skip a few pages).

Of course, the concept of intelligence itself is a difficult one to pin down and quantify. IQ tests reflect only one aspect of intelligence, the one which seems most influential in gaining material success in Western societies. In fact, different components of brain function— memory, association, extrapolation, intuition, and creativity—are all important, working separately and together. IQ tests only measure limited aspects of these diverse talents, and there is no doubt that cultural differences and familiarity with the contents of tests affect results. For example, Native Americans generally get very low IQ scores, even though they were originally Orientals, the alleged superiors of Europeans and Africans.

Unfortunately these issues have generally been overlooked in the furious babble that greeted publication of *The Bell Curve* and sister works. Particularly ironic was the coverage in *The Times* (London) which, on a day when it carried some of the very first reports of the debate, published an obituary of Davidson Nicol, a black African from Sierra Leone who got a first in Natural Sciences at Cambridge,

then gained a medical degree, and finally followed this up with a distinguished academic and diplomatic career.[19] His story in itself might
have seemed a rare or isolated accomplishment. Yet only three
months later *The Sunday Times* (London) reported that:

> Black Africans have emerged as the most highly educated
> members of British Society and are twice as likely to hold jobs
> in the professions as white people. The findings, which chal-
> lenge popular stereotypes about black underachievement, are
> revealed in a new analysis of the official census that for the first
> time, details the social class, educational achievement and aspi-
> rations of Britain's three million ethnic minorities. More than a
> quarter of the 130,000 adult black Africans in Britain hold
> qualifications higher than A-levels, compared with one in eight
> whites. They are now just ahead of the Chinese, the most aca-
> demically successful ethnic minority in previous studies.[20]

Yet the authors of *The Bell Curve*, and all its apologists, argue that
putting resources into education to improve the intellectual performance of children from deprived backgrounds is a waste of money. It
is a view denounced by Tim Beardsley, writing in *Scientific American*:

> Even though boosting IQ scores may be difficult and expensive,
> providing education can help individuals in other ways. That
> fact, not IQ scores, is what policy should be concerned with.
> *The Bell Curve*'s fixation on IQ as the best statistical predictor
> of a life's fortunes is a myopic one. Science does not deny the
> benefits of a nurturing environment and a helping hand.[21]

In any case, the story of our African Exodus makes it unlikely
there are significant structural or functional differences between the
brains of the world's various peoples. We came out of Africa as an
already advanced species and those who remained on the continent
retained that sophistication, just as much as the rest of *Homo sapiens*
used it to conquer the world. Of course, that does not mean there
are absolutely no variations between populations. As we have seen

already, the "race" called *afer* by Linnaeus, Ethiopian by Blumen-bach, Congoid by Coon, and Negroid or African by Rushton appears to contain as much genetic variation as the rest of humanity put together, a fascinating prospect that will be put to the test over the next few years as the Human Genome Diversity Project gathers data from around the world.[22] This venture—an offshoot of the Human Genome Project which is scheduled to produce a composite, but complete, map of the entire human genetic code by the early years of the next century—will study certain stretches of our DNA to see how they vary among different peoples. The results will provide a new, far more realistic perspective on racial differences. As Howard University's Georgia Dunston puts it: "After the diversity project we won't have the luxury of drawing distinctions between one another based on skin pigmentation."[23]

It is a point backed by Jared Diamond:

Of all those traits that are useful for classifying human races, some serve to enhance survival, some to enhance sexual selec-tion, while some serve no function at all. The traits we tradition-ally use are ones subject to sexual selection, which is not really surprising. These traits are not only visible at a distance but also highly variable; that's why they became the ones used throughout recorded history to make quick judgments about people. Racial classification did not come from science but from the bodies' signals for differentiating attractive from unattrac-tive sex partners, and for differentiating friend from foe.[24]

In other words, we have extenuated the minute differences between ourselves, sometimes with grievous results.

Indeed, says Loring Brace, the concept of race is probably only a very recent one. For most of our existence we lived without the notion and without a term for it in our vocabularies, he believes. Only when we came to the great age of European exploration which began in the fifteenth century did one set of people encounter another that looked starkly different. "The concept of race did not exist until the invention of oceangoing transport in the Renaissance," he states in an article in *Discover* magazine. Before then explorers traveled on horseback, covering only about twenty-five miles a day:

> It never occurred to them to categorize people, because they had seen everything in between. That changed when you could get into a boat, sail for months, and wind up on a different continent entirely. When you got off, boy, did everybody look different! Our traditional racial groupings are not definitive types of people. They are simply the end points of the old mercantile trade networks.[25]

Ideally, we need a time machine to travel back into our recent— and distant—past to prove such points. It would be an enriching exercise: most illuminating of all, would be that period—about 40–60,000 years ago—when small and separate human populations started to expand, marking the beginning of our planetary takeover. Those surges in numbers acted like biological photocopiers creating multiple versions of these newly divergent people of different shapes and hues. After that, natural selection, sexual selection, and isolation helped mold our species in increasingly diverse ways. Later, over the past 15,000 years, as the fingers of ice that covered earth's higher latitudes relinquished their grip, the boundaries between different populations became blurred as benign climates and swelling human ambitions further mixed the melting pot of populations.

And that is the task of the Human Genome Diversity Project. Its unraveling of our planet's web of ancient lineages will be a fascinating business, though understanding what brought about our global spread is a different question. Obviously there was something very special about the mind of *Homo sapiens* that led to our African takeover and exodus, an evolutionary endowment that dates from the creation of our species more than 100,000 years ago. That gift was present in that small founding population and it gave us all a shared heritage of social intelligence which was one of the keys to our success. That mental capacity was (and is) extraordinarily complex and if we want to understand our own basic nature, we would be best occupied in extending our understanding of its general nature rather than attempting to find minute variations in human ability in order to use them as the grounds for discrimination. Ever more sophisticated methods of probing the workings of the brain are being developed, and these will allow us to move from crude generalizations to detailed studies of its operations.[26] For instance, it has been claimed that functional differences between male and female brains have been found. But even if these exist, they probably have a much more ancient basis in behavioral variation between the sexes (see Chapter 8) than the short timescale of differentiation which separates the human races.

This brings us directly to the final mysteries surrounding our African Exodus. What were the secrets which allowed us to grow in numbers and to spread to new lands previously beyond our reach? What finally propelled us down the road to today's technology? This is an elusive topic, of course. In Chapter 4, we showed how superior organization may have given our ancestors a crucial lead over our hominid cousins, the Neanderthals, an advantage that may also have helped us to compete with archaic people in China and Java. In addition there was our undoubted technological expertise—our mastery of boat building, working skins into clothes, constructing better houses and hearths, and other skills—that opened up previously untameable terrain.

But what biological, behavioral, or cultural blessing underpinned these achievements, and what was its basis? What evolutionary pressures drove *Homo sapiens* toward these changes? These are critical issues, which have had an enormous impact on our lives, and which we shall now examine.

8

The Sorcerer

Intelligent life on a planet comes of age when it first works out the reason for its own existence.

Richard Dawkins

Imagine the following scene: a group of youths, their bodies plastered with red ocher and wrapped in carefully-sewn fur clothes, are led through a maze of underground chambers by their tribal elders. Deep inside the cavern, they are kept waiting in a small cave, huddled in the dark. Finally they are ushered into a vast rock sanctum lined with lurid animal paintings and lit by the guttering flames of oil lamps. A horse, painted in black, rears up out of the bulging limestone wall, dancing in the flickering light. A trio of lions stare menacingly into space. Over an archway, a strange figure—part human, bird, and deer—prances in the gloom. Drums are pounding, there is chanting, and the cave is filled with thick black smoke.

The effect, of course, is terrifying—intentionally. Their day in "the gallery of the beasts" will be etched forever in those young minds, bonding them together, and to their tribe. Perhaps the ritual was male-dominated and linked to hunting, the creatures on the wall representing prey, or possibly revealing qualities sought by human hunters. Alternatively, ceremonies celebrating female sexual maturity may have dominated proceedings. Either way, those magnificent red and black images of bison, rhinos, ibex, cattle, and deer, would have been of critical importance in cementing the edifice of tribal life

49   Copies of images from Cro-Magnon cave art.

20,000 years ago. Today, we see the remnants of that social reinforce-
ment slowly gathering calcite on the walls of Altamira, Lascaux,
Vallon-Pont-d'Arc, and all those other caves once favored by Cro-
Magnon society.

The artwork is, of course, superb and those responsible have
rightly been hailed for their sophistication and talent. However, the
figures they drew and colored—possibly by a sort of oral spraying, in
which pigments, such as ocher, blood, and soot, mixed with saliva,
were blown onto walls—are more than mere demonstrations of early
aesthetic sensibility. They may be manifestations of the last, critical
step in the flowering of human intellect itself, the final additions to
our mental architecture that took us from being clever hominid toolmakers to masters of the planet—though the emergence of these features is not quite the straightforward business that some scientists
make out, as we shall see.

Certainly, something very special was happening to human society
around this time. Before then, *Homo sapiens* was simply marking time
culturally. For millennia upon millennia, we had been churning out

the same forms of stone utensils, for example. But about 40,000 years ago, a perceptible shift in our handiwork took place. Throughout the Old World, tool kits leapt in sophistication with the appearance of Upper Paleolithic style implements. Signs of the use of ropes, bone spear points, fishhooks, and harpoons emerge, along with the sudden manifestations of sculptures, paintings, and musical instruments. As John Pfeiffer states in *The Creative Explosion*: "Art came, with a bang as far as the archaeological record is concerned."[1] We also find evidence of the first long-distance exchange of stones and beads. Objects made of mammal bones and ivory, antlers, marine and freshwater shells, fossil coral, limestone, schist, steatite, jet, lignite, hematite, and pyrite were manufactured. Materials were chosen with extraordinary care: some originated hundreds of miles from their point of manufacture. In Europe, only a dozen or so of the thousands of shell species available on the Atlantic and Mediterranean shores were turned into ornaments, while only the teeth of certain animals were chosen as a raw material. On a more utilitarian level, storage pits, huts (including some found in the Ukraine made entirely of mammoth bone), and mineral quarries were crafted—to be followed, eventually, by the domestication of animals and plants, the beginning of metallurgy, and our move down the road to the Roman Empire, the Renaissance, nuclear energy, and the wonders of space travel.

It is an extraordinary catalogue of achievements that seem to have come about virtually from nowhere—though obviously they did have a source. The question is: What was it? Did we bring the seeds of this mental revolution with us when we began our African Exodus, though its effects were so subtle they took another 50,000 years to accumulate before snowballing into a cultural and technological avalanche that now threatens to engulf *Homo sapiens*? Or did that final change occur later, and was it therefore more profound, and much speedier in its effects?

Many archeologists, linguists, anthropologists, and researchers in other fields, have little problem over their preferred response. Only the former makes sense to them, though this acceptance has implications bristling with intellectual difficulties. Their reasons for disavowing the latter scenario are simple. If we accept that neurological

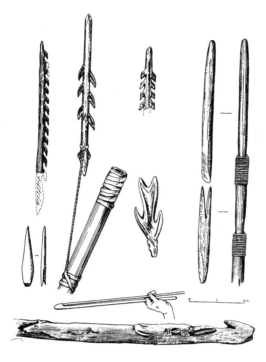

50   Cro-Magnon bone and antler tools, and the use of a spear-thrower.

or behavioral changes were responsible for the abrupt flowering of human culture only about 40,000 years ago then we have to explain how this moment of transformation occurred more or less simultaneously across Africa, Asia, and Europe. As we have seen, the DNA studies by Harpending, Rogers, and others demonstrate that a whole variety of people—Turks, Sardinians, Australians, Japanese, Native Americans, and others—all went through sudden, and rapid eruptions in populations at the same time that our mystical Cro-Magnon creative explosion was occurring.

Clear evidence for rises in human numbers and in artistic and technological sophistication has only been found in Europe, but then—paleontologically speaking—this is the world's most assiduously studied continent. Elsewhere, research has been patchy and inconclusive until recent tantalizing evidence first emerged to show other parts of the Old World were also going through the rigors of artistic and cultural upheaval; a social tumult which matches those peaks of mitochondrial DNA mutations detected by Harpending and the rest. We now know that *Homo sapiens* sailed boats from southeast Asia to Australia at least 50,000 years ago and although it is unlikely these hapless settlers were actually aiming for the land in question, the fact that they were nautically mobile in the first place indicates they possessed considerable sophistication. Then there was the use of red ocher, the practice of cremation, and the creation of paintings, engravings, and necklaces—all discovered in Australian sites that are 30,000 years old or more. Intellectual change was in the air, and was not restricted to one small part of one continent. It was blossoming across the populated world.

But how can we rationalize the fact that artistic and symbolic ferment was bubbling from Mungo to Lascaux? If we assume the requisite cognitive transformations were triggered after *Homo sapiens* embarked on its African Exodus, we must explain how it erupted in tribes in different parts of the world. If the Great Change occurred as late as about 40,000 years ago, then it must have developed, virtually simultaneously, in peoples who were living many thousands of miles apart. Either that or those new genes or behaviors appeared in one place and then spread like social wildfire across half the planet, a notion that can only be sustained by assuming our forebears were indulging in a lather of gene or cultural exchange. This has led many scientists to see it as improbable that this decisive event occurred so late in our prehistory, when, they say, all sorts of convoluted, tortured explanations must be dreamed up to account for its transmission.

Instead, these scientists argue that mental mutations, which had evolved much earlier, more than 100,000 years ago, when *Homo sapiens* was still confined to one small part of its African homeland, were responsible for later artistic and technological development. This decisive modification was taken round the world with those humans who embarked on our great exodus. (It also remained with those who stayed in Africa, of course.) Such an explanation is unaffected by a need to have mysterious flows of DNA or cultural mutations striking most of the world's peoples simultaneously. On the other hand, if neuronal remodeling was part of our African heritage, why did it take so long to manifest itself? Why, when we look at those populations at Amud, Skhul, Qafzeh, Kebara, and the other Levant sites, can we see precious little difference between the stone tools made by *Homo sapiens* and our Neanderthal cousins? If we were already bequeathed with our full, final neurological endowment, why did it not manifest itself then? We had begun our exodus, and by inference were kitted out with our complete intellectual baggage. We were, to all intents, the same sorts of people that we now encounter every day at home or work. So why did our sophistication not show then? Why the delay before it became apparent?

These are vexing questions whose answers depend on understanding not just the hardwiring of our brain, but the evolution of our social structures, artistic needs, technological prowess, and much else. All these features interacted as we unfolded as a species, which means we have to look at a broad biological canvas before we can understand what was so special about *Homo sapiens*, and to appreciate why it carried such a powerful neurological blessing—one that may nevertheless have taken 60,000 years to take effect. It is an examination which produces no clear answers, of course. "The long, tedious, and relatively unchanging patterns of the Middle and particularly the Lower Palaeolithic contrast so dramatically with more recent remains that a disjunction is indicated when one looks ahead from the past," says American anthropologist Lew Binford. "Looking back from the present, however, we have generally failed to seek a processual understanding of this transitional event."[2] In short, science has not yet made up its mind how to resolve this conundrum, though researchers do have their ideas, based on a new approach that is being taken to the study of human behavior. This is the science of evolutionary psychology, whose tenets are based on the assumption that we are not biological devices newly constructed for the twentieth century, but carry the trappings of our recent Stone Age origins in our minds (and our bodies as we shall see in the next, and final, chapter of this book).

Now our evolution, as we have seen, can be traced back to those swings in climate that swept Africa over the past few million years, and in these variations we can track the founding of human intelligence. "The evolution of anatomical adaptations in the hominids could not have kept pace with these abrupt climate changes, which would have occurred within the lifetime of single individuals," says neurophysiologist William Calvin, of the University of Washington School of Medicine. "Still, these environmental fluctuations could have promoted the incremental accumulation of mental abilities that conferred greater behavioural flexibility."[3]

In other words, our bodies could not change speedily enough, so our brains took the strain instead. We developed a plastic, adaptive approach to the world. The result was a doubling in the expansion of our crania, a process which began around two million years ago, when *Homo habilis* and then *Homo erectus* people started to gather round the lakes of eastern Africa to make their tools and plan their scavenging and foraging (and possibly hunting). Their brains had, roughly, the capacity of a pint pot. Then, slowly, we began to gain gray matter, at a rate of about two tablespoons' worth every 100,000 years. By the time this cerebral topping-up had finished, the human cortex had more than doubled in volume. "The two-millimeter-thick cerebral cortex is the part of the brain most involved with making novel associations," adds Calvin. "Ours is extensively wrinkled, but were it flattened, it would occupy four sheets of typing paper." In fact, this outer layer of gray matter accounts for about 80 percent of our total brain volume. Compared with a human's, a chimpanzee's cortex would fit on only one sheet of paper, a monkey's on a postcard, a rat's on a stamp.

The contrast is helpful, though taking a rolling pin to the brain does not tell you much else about the evolution of human "genius." For that we need to look elsewhere, not at flattened neural wiring. In fact, the proper pursuit is through asking the right sort of questions. In this case, we want to know: What is the brain for? It is a point stressed by psychologist Leda Cosmides, of the University of California, Santa Barbara:

If you were an alien confronted by a toaster for the first time, you would be very puzzled about its purpose. You could take it to pieces and learn how it works. But no amount of dismantling would tell what it was for: burning bits of bread to make toast. We should ask the same sorts of questions about the brain and the mind. People have not done that properly in the past.[4]

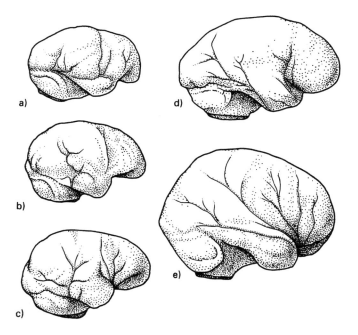

51 Casts of the brain cavities of chimpanzee (a), gracile australopithecine (b), robust australopithecine (c), *Homo erectus* (d), and modern human (e).

For most of the history of psychology, queries about the brain have centered, not on its purpose, but on what it can do, an approach that has generated metaphors comparing our minds with blank slates, ready to be drawn on, or computers that can be programed in response to external stimuli. In either case, it is assumed our minds are empty vessels which are filled in ways specified by our culture, thus implying that we get our language, fear of spiders, appreciation of beauty, gender identity, aversion to incest, desire to have friends, and countless other thoughts and feelings from our surroundings. Without that input, we would be intellectual voids. Now this is a strange notion when viewed from the perspective of a book such as this one, in which we have sought to make sense of the human condition by studying our rise from primate tree-dwellers to modern technocrats. It is a story which highlights our minds' increasingly effective power to influence our surroundings—and not the other way round. This is not to say that culture has not been critical to the rise of *Homo sapiens*. This is certainly not the case, as we can see on occasions when its potency has been weakened, for instance, when Tasmanian Aborigines became isolated from their mainland cousins 10,000 years ago and lost some of their technological sophistication. Human brains have generated communal creative milieus that have enhanced intellects, a "bootstrap process" in which the first such adaptation engendered profound behavioral changes which, in turn, made larger cortexes progressively more advantageous. This positive feedback loop propelled the evolution of increased brain size towards that of *Homo sapiens*. "The melding of genetic change with cultural history, both created the mind and drove the growth of the brain and the human intellect forward at a rate perhaps unprecedented for any organ in the history of life," state Charles Lumsden and Edward O. Wilson in their book, *Promethean Fire: Reflections on the Origins of Mind*.[5] The primary source of this mental leap was still the human brain, however. Culture came later.

It is a point stressed by Leda Cosmides, and her colleague (and husband) John Tooby, also of the University of California, Santa Barbara:

Knowing that the circuitry of the human mind was designed by the evolutionary process tells us something centrally illuminating: that, aside from those properties acquired by chance or imposed by engineering constraints, the mind consists of a set of information-processing adaptations, designed to solve those problems that our hunter-gatherer ancestors faced generation after generation. The better we understand the evolutionary process, adaptive problems and ancestral life, the more intelligently we can explore and map the intricacies of the human mind.[6]

Such a strategy forms the core of evolutionary psychology which tries to examine our conduct from the perspective of a hunter-gatherer with five million years of hard hominid evolution to its credit, and whose occasionally baffling actions and reactions can be best understood in this light. It is an approach that accepts a certain programed response from the human mind, but it does not maintain an individual is necessarily a prisoner of his or her genetic heritage.

Our evolution has been too complex for that. As Robert Wright, author of *The Moral Animal: Why We Are the Way We Are*,[7] points out: "What is 'natural' is not necessarily unchangeable. Evolutionary psychology, unlike past gene-centred views of human nature, illuminates the tremendous flexibility of the human mind and the powerful role of environment in shaping behaviour." This tactic has two key advantages. It places great stress on our evolutionary history, thereby helping us to understand *Homo sapiens*' special mental demeanor and it offers us realistic prospects of finding solutions to the mystery of our brains' last neurological changes. We shall look at these shortly, but before we do we should also mention Darwinian psychology's other substantial benefit: that of suggesting insights, available through no other route, into human nature today. Why do we erupt in violence on certain occasions? Why do we choose the mates we do? How are we able to live in large groups, some containing millions of individuals living in cramped, close proximity? Of course, our environments have an enormous impact on our behavior. Nevertheless, there are some appealing intuitions offered by evolutionary psychology that other branches of the science simply cannot manage. Indeed, with the benefit of hindsight, it seems quite extraordinary that it has taken the profession so long to appreciate the paleoanthropological point.

However, first let us return to that vexing problem of the Great Mental Leap.

What evolutionary psychology teaches us is that our ancestors must have had to evolve a whole series of mental mechanisms—those "sets of information-processing adaptations" mentioned by Cosmides and Tooby—that were used to solve the problems of everyday Stone Age life: food selection, mate selection, communications, toolmaking, dealing with wild animals, and many more. "Think of the mind as a great Swiss Army knife," says Cosmides. "We had different mental blades for solving all sorts of problems."[8]

More than any other species, including our reconstructions of other hominids, *Homo sapiens* possesses a wide array of different mental tools that we use for dealing with the outside world. And because we have such a variety, we can react more flexibly and deal with issues which we would never have encountered in our evolution. "It is like that Swiss Army knife," adds Cosmides:

I have a screwdriver on mine, but I could also use it as a hole puncher. It is not what it was designed for, but I can still exploit it that way. Similarly with our minds. Take the example of learning to speak and write. Now spoken language acquisition is instinctive but not writing. People learn the former by simply listening, we have to be taught the second. And that is because we have learned, only recently, to use our language acquisition skills, combined with some input from our vision to make marks that convey our words, i.e., writing. We have used our hunter-gatherer skills to do something radically new.

So let us look at some of the blades of this mighty neurological knife to see if we can find clues to the source of our species' global success. And of these, there is one that cries out for particular attention: language. For Darwin, language was "an instinctive tendency to acquire an art,"[9] and so the subject was treated in Victorian days—as an art. Today, however, its study is very much the province of the scientist, from the neurophysician to the computer expert. To them, or at least most of them, language is a hereditary endowment, as Steve Pinker stresses. "Language ... has been found in every one of the thousands of societies that have been documented by explorers and anthropologists," he says. "Within a society, all neurologically normal people command complex language, regardless of schooling."[10]

We are, in fact, a startlingly talkative species, so much so that the British phonetician D. B. Fry has remarked, tongue in cheek (but then he is a phonetician) that *Homo loquens* would be a far more appropriate name for our species than *Homo sapiens*.[11] Certainly, our urge to communicate is generally more in evidence than our wisdom. It is estimated that in a normal day, a person may speak as many as 40,000 words, the equivalent of four to six hours of continuous speech. And most of these are expended debating issues that could not be categorized as the products of a particularly sapient, i.e., wise, animal. Professor Robin Dunbar, of Liverpool University, has analyzed conversations in several universities' common rooms and found that only 14 percent had anything to do with academic matters. Most talk was "trivial" as Dunbar puts it, dwelling on personal relationships and experiences. "Culture, science, and religion, even sport, account for a surprisingly small proportion of conversation topics," he says. On the other hand, gossip, in which we exchange information about each other, our love lives, and the plots of sundry TV soaps, accounts for about 70 percent of chat time. "It's what makes the world go round," adds Dunbar.[12]

The crucial point is that language is ubiquitous among humans, a facility that is acquired just from exposure to the speech of the people with whom children interact. It is a facet of the mind quite separate from general intelligence, for language can be handicapped even though intelligence is left intact—and vice versa. More importantly, it is a means of communication bursting with extraordinary evolutionary implications. "In an intelligent social species such as ours, there is an obvious adaptive benefit in being able to convey an infinite number of precisely structured thoughts merely by modulating exhaled breath," says Pinker. "Anyone can benefit from the strokes of genius, lucky accidents and trial-and-error wisdom accumulated by anyone else, present or past."[13]

In other words, it was the genetic capacity to speak a complex language that raised modern humans from the millennia-long doldrums we were sharing with the Neanderthals until 40,000 years ago. It gave us the power to take over the world. This interpretation is shared by others. "Our success must have had a lot to do with speech which is, after all, an enormously complex process," says Professor Kidd, of Yale's human genetics department:

When we talk, more than 100,000 neuromuscular events are triggered every second, and the movements of more than 100 muscles have to be co-ordinated. The diaphragm, tongue, cheeks, and jaw all have to be controlled. That whole process is extraordinarily difficult. For humans, this was a triumph of evolution and it set us apart from the rest of the animal kingdom.[14]

Armed with the power of speech, humans would have been able to describe precisely where fruit and vegetables were growing, direct elaborate hunts, and allow tribal elders to recount how famines had been conquered. Other hominids, such as the Neanderthals, may have spoken, but it would have been cruder, and less effective compared with the sophisticated language of modern humans, say proponents.

One such adherent is Jane Goodall, whose chimpanzee studies have revolutionized our understanding of our nearest primate kin and have, in the process, shone a great deal of light on our own foibles and talents. Her remarks about the importance of language are particularly striking. She writes:

> Of all the characteristics that differentiate humans from their non-human cousins, the ability to communicate through the use of a sophisticated spoken language is, I believe, the most significant. Once our ancestors had acquired that powerful tool, they could discuss events that had happened in the past and make complex contingency plans for both the near and the distant future. They could teach their children by explaining things without the need to demonstrate. Words give substance to thoughts and ideas that, unexpressed, might have remained, for ever, vague and without practical value. The interaction of mind with mind broadened ideas and sharpened concepts. Sometimes, when watching the chimpanzees, I have felt that, because they have no human-like language, they are trapped within themselves. Their calls, postures and gestures, together, add up to a rich repertoire, a complex and sophisticated method of communication. But it is non-verbal. How much more they might accomplish if only they could talk to each other.[15]

Goodall's view is supported by others, such as Calvin. "Language is the most defining feature of human intelligence: without syntax—the orderly arrangement of verbal ideas—we would be little more clever than chimpanzees," he says. The notion that *Homo sapiens'* linguistic prowess is of a special quality is also supported by the work of Philip

Lieberman (father of Dan, whose Neanderthal and armadillo bone research we encountered in Chapter 4), Jeffrey Laitman (whose work on Neanderthal anatomy was also described in that chapter), and others.[16] They have studied the skull bases of apes, humans, and extinct hominids, and concluded that the more arching of this section, the more language competence was possessed by its owner. And Neanderthals, they found, scored noticeably poorer results than *Homo sapiens* with regard to skull base arching, suggesting they were simply less able to communicate than our immediate ancestors were.

However, these ideas are countered by the work of Yoel Rak, Bernard Vandermeersch, and others.[17] At a dig at Kebara, they found the finely preserved 60,000-year-old skeleton of a Neanderthal containing a complete hyoid bone (also mentioned in Chapter 4). The hyoid acts as an anchor point for the vocal tract in our throat, and if the Neanderthals were noticeably poorer speakers we might expect to see signs that their versions are differently shaped from ours in some way. We do not. The Kebara hyoid is identical to a modern human's—providing no evidence that Neanderthals could not speak as well as we can today. In other words, it may not have been better articulation that did the trick for *Homo sapiens*. However, the source of our mental "leap" could still have lain with our language. The complexity of the ideas imparted by our speech would have depended, not on our vocal tracts, but on our brains. In other words, it was the content, not the form, that mattered.

Or perhaps we needed brains of increased size for a different, but equally rudimentary reason: to weld larger numbers of *Homo sapiens* into bigger bands of hunter-gatherers, using language as social "glue." In other words, complex speech was subordinated to a secondary cause—to hold together substantial, complex groups of humans. Primates are particularly sociable creatures, of course, and use grooming—the laborious picking through fellow troop members' fur for extraneous items—to establish social positions and create liaisons. However, there is a ceiling to how much grooming an individual animal can do, without encroaching on other vital activities such as feeding. That limit in turn restricts the numbers of possible alliances it can establish, and ultimately puts an upper limit on group size. Complex speech allowed us to overcome this problem by carrying out the equivalent of several acts of grooming simultaneously. And so numbers in bands of humans jumped, says Dunbar. This notion implies that language did not evolve as a method for relaying data, but for more subtle purposes, as Pinker stresses. "Human communication is not just a transfer of information like two fax machines connected by a wire; it is a series of alternating displays of behavior by sensitive, scheming, second-guessing, social animals."[18]

The idea also dovetails nicely with the evidence that social matters (i.e., gossip) dominates human verbal communication, as we saw earlier in this chapter. All that time we spend chatting about social slights, hangovers, rows, and TV plots in bars, corridors, and common rooms is the equivalent of the hours gibbons, or chimpanzees, invest in picking through group members' fur in order to establish their social positions. A primate species that can expand its grooming capacity can enlarge its group size and effectiveness. Language fulfills that requirement and can be carried on along with activities, unlike grooming, which occupies the hands and eyes. More liaisons became possible, and bigger, more cohesive, bands of people in their wake. "Speech enables us to exchange information about each other, and so greatly speed up the rate at which we learn about our constantly changing social universe," says Dunbar. "That, in turn, helps to ensure that the group remains cohesive." And do not forget the evidence revealed in Chapter 4 in which we showed how improved social intelligence, and wider webs of trade and exchange characterized the behavior of Cro-Magnons compared with Neanderthals.

So far so good. But is there any hard evidence to support this ingenious thesis? Well, yes, there is, says Dunbar. If you look at primate brain dimensions, you find that they correlate neatly with group size, he points out. Gibbons have fairly small crania and live in family pairs of four to six, for example. Their neocortexes (the most recently evolved parts of the cortex) contrast with that of bigger-brained chimps who live in communities of fifty to eighty. The relationship is unusually clear-cut, he maintains. And when you plug the human brain size into this social thermometer, you produce a predicted group size of 148—a figure that is the optimal maximum for social assemblies of humans. Amazingly, says Dunbar, this magic number turns up in all sorts of human societies. Many hunter-gatherers alive today have an average core group of around 150, and so did Neolithic villages, such as those uncovered in Mesopotamia. In addition, the company—the smallest unit that operates as an independent group in an army—hovers around this number in most countries, 135 for Britain to around 200 for the United States. And then there was Brigham Young, who, when organizing the great Mormon trek from Illinois to Salt Lake City, divided his 5,000 followers into groups of 150. In short, the number looks like the fundamental unit of human social cohesiveness. Above this level, peer pressure can no longer control individuals and the group breaks apart. By using language to create this largest of all primate or hominid assemblies, *Homo sapiens* was able to generate a healthier, more effective culture.

It is an intriguing idea, though not everyone agrees that speech—either as a primary or secondary mental product—is necessarily the final deciding factor in humanity's success story. The gift of the gab may have taken us far down our present evolutionary track, but it was not necessarily the final means of our current "successful" status. For this reason, some scientists champion the cause of different "brain blades," such as memory. The storage of extra neural information would have been of immense benefit, they say. There would have been no point in having language if we did not have the power to retain the complicated knowledge that we wished to pass on, after all. With good memories we would have been able to sustain complex social relations. We could recall where we saw good hunting grounds the previous year and where we could find food supplies and vegetation. Tied to this notion is the issue of longevity. If humans lived, on average, to an older age, we would have been able to pass on more cumulative wisdom stored in our memory banks. There would have been more elders to transmit the benefit of their knowledge: what had been done in their youths during serious drought, for example. In other words, it was the rise of the human grandparent that gave our species its precious boost.

Then there is the idea, backed by Binford, that *Homo sapiens* pos-

sessed—but Neanderthals lacked—the genes that control the neurological power to plan in depth, an ability that allows us to foresee and
plan for alterations in our circumstances and to map and fully exploit
resources. Binford points out that modern hunter-gatherers often
initiate actions long before their anticipated need has manifested
itself. He points out that:

> A move to a fish camp along a salmon stream is generally made
> before salmon appear in the stream, on the basis of stored and
> analysed knowledge of the environment and of the behaviour of
> fish. The group may well engage in the manufacture and repair
> of fishing gear long before any indirect indication that salmon
> are present, will be present, or might be exploited. When the
> salmon arrive, heavy labour investments are made in obtaining
> fish, which are then processed for stores that may serve as food
> for the group over a six-to-eight-month period.[19]

Such behavior typifies *Homo sapiens*, and delineates our behavior
more than any other hominid, argues Binford.

Alternatively, it may have been a deeper underlying mental structure that went through a final crucial change, Calvin suggests:

> To account for the breadth of our abilities, we need to look at
> improvements in common-core facilities. Environments that
> give the musically gifted evolutionary advantage over the tone
> deaf are difficult to imagine, but there are multifunctional brain
> mechanisms whose improvement for one critical function might
> incidentally aid other functions. We humans certainly have a
> passion for stringing things together: words into sentences,
> notes into melodies, steps into dances, narratives into games
> with rules of procedure. Might stringing things together be a
> core facility of the brain, one commonly useful to languages,
> storytelling, planning, games and ethics? If so, natural selection
> for any of these talents might augment their shared neural
> machinery, so that an improved knack of syntactical sentences
> would automatically expand planning abilities too.

Now the notion that only a slight difference in any one of these mental attributes (memory, language, planning ability, etc.) might have produced such a startling variation in outcome—world domination for us, extinction for the Neanderthal, Ngandong, and Dali peoples and the rest—may seem improbable. Yet it is clear that only a relatively minute change in genes and behavior could have accounted for our disparate fates. Although they looked dissimilar with regard to forehead angle, browridge size, and other features, these people—underneath their skins—must have been very like *Homo sapiens*, as a casual calculation indicates. Chimps and humans diverged five million years ago and our genomes only differ by about 2 percent, according to DNA hybridization studies. However, the separation between Neanderthals and modern humans probably occurred only about 200,000 years ago which suggests that we may have differed by less than 0.1 percent of our genomes. And that slender gap may account for our success and their failure. Only a handful of genes must be involved in our jump to stardom, it would seem.

In which case, says Steven Mithen, of Reading University, it may not be a question of which blade of our Swiss Army knife was honed to final perfection, but more the way we integrated their use most effectively. There was no big gap in individual mental aptitudes, just a different way of putting them together.

Mithen categorizes those mental knife blades into different "domains": including those which had a social function, such as language; those which had a technical function, such as the skills involved in toolmaking; and those to do with "natural history intelligence," the knowledge we acquired about our environment, and its resources, including the animals we hunted. "Early humans appear to have been unable to integrate their thought and knowledge from these multiple cognitive domains," Mithen says:

For example, Neanderthals were under severe adaptive stress— 95 percent of them were probably dead by the age of thirty-five. In such situations, it would seem to have made great ecological sense to have applied their technical skills to making beads and pendants to facilitate social interaction, or to have improved foraging efficiency. But they didn't. They had a domain-specific mentality: not for them the confusion and conflation of aims and criteria, but a clear sightedness, and a single mindedness absent from the modern mind.[20]

In a sense, the Neanderthal mind represented the culmination of millions of years of development of specialized intelligence in the primate line. With modern humans, it stopped, and we became much more fluid and generalized in our thought processes. Ideas crossed over from one intellectual domain to another. Language, which had been primarily concerned with social interaction, was used to communicate information about all sorts of things: toolmaking technology, the natural world, and much more. The boundaries between social and nonsocial behavior would have become fuzzy—as they are with modern hunter-gatherers. "Consider, for instance, attitudes to the natural world," says Mithen. "It is ubiquitous among forest-dwelling hunter-gatherers to conceive of the forest as parents, it is in effect a social being that gives. Similarly the Inuit living in the Canadian Arctic view their environment as imbued with human qualities of will and purpose." The consequence of this cross-fertilization was the birth of anthropomorphism and totemism, a belief in kinship with the animal world. To a Neanderthal, a cave bear was a cave bear. To a modern human, it was not only a threat, or possibly a source of food, it was a god, an ancestor, and who knows what else.

We see signs of this intermingling of ideas and intellectual domains in those cave paintings, such as the strange figure—part human, bird, and deer—that was highlighted in this chapter's opening paragraphs. Although our cavern was a hypothetical one based on elements garnered from several of the two hundred or so that have been discovered to date, that strange totemic being is a specific one: it is The Sorcerer, a painting from Les Trois Frères cave in the Pyrenean foothills of Ariège (named after the three sons of Count Begouen who discovered it in 1912). This is a therianthrope—a part human, part animal, part mystic being—that is the creation of an astonishing imagination. "The body is uncertain, but is some kind of large animal," says Denis Vialou, of l'Institut de Paléontologie Humaine, Paris. "The hind legs are human, until above the knees. The tail is some kind of canid, a wolf or a fox. The front legs are abnormal, with humanlike hands. The face is a bird's face, with deer's antlers."[21] Even more striking is the way this magical creature gazes directly out of the wall in a stare that pierces the attention of the onlooker.

Quite clearly, the Cro-Magnons were able to fuse the natural and social worlds in a highly imaginative fashion. We see the manifestation of this prowess in an even earlier work (The Sorcerer has been dated as being about 15,000 years old). This is the 32,000-year-old Lion Man, the earliest known therianthropic figure, a carved mammoth ivory shape, kept at the Museum der Stadt, in Ulm, Germany, that was originally thought only to be that of a body of a human.[22] One day, curators were presented with a lion's head, also of ivory. At first, they saw no connection. Then, one remembered the ivory body of the human. They put them together—and found they fitted perfectly. The foot-long statuette may represent a shaman, or sorcerer, but clearly indicates a very early symbolic use of the intermingling of different intellectual

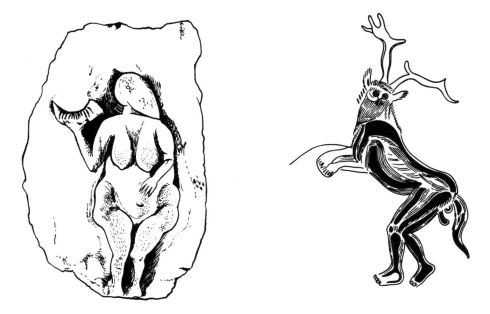

52   Two famous images from Cro-Magnon art: A "Venus" from Laussel, and "The Sorcerer" from Trois Frères.

domains: the natural (the lion); the social (the human figure); and the technical (the act of carving).

We should not forget, however, that for all the fact that these were inventive times, they were also extremely oppressive. In Europe, modern humans had to endure some of the worst climatic vicissitudes ever experienced by *Homo sapiens* as the last Ice Age slowly ground life to a virtual halt. By 20,000 years ago, glaciers covered the entire northern half of the continent with separate sheets of ice stretching down from the Pyrenees and Alps. Nevertheless, in the midst of this icy desolation, in the Dordogne and Ardèche, thrived the Cro-Magnon—a recent émigré from Africa who was then enduring a climate like that of modern Iceland or Greenland. "And the only way they could have managed that was through sophisticated interaction," says archeologist Professor Clive Gamble of Southampton University. "There would have been no room for individual iconoclasts then. Lack of co-operation meant death."[23]

The fact that Cro-Magnons developed an expertise as consummate artists, in the midst of this glacial grimness, was no coincidence. Art—generated through the merging of mental domains, and the breaking down of intellectual barriers—must have been critical to the business of survival. We can see this in the fact that people then baked statuettes of clay: demonstrating that Cro-Magnons had the technology to make pottery but chose only to make sculptures. In other words, symbolism and art were as important as functional applications in those times. "Art had a social role in the beginning," says Gamble. "It was used to solve the problem of who belonged where, and of what roles a person had to fulfill in order to ensure groups and tribes could survive a tight, hard way of life. Our predecessors needed to have that constantly reinforced." The roots of art lay with a need to create initiation ceremonies, to hold rituals, to settle territorial squabbles, and to demarcate roles in society, such as hunting. It was all part of the power we developed to free ourselves from the strictures of individual intellectual realms. If there was one thing that identified *Homo sapiens* at this time, it was mental osmosis in which ideas would creep throughout the mind to produce the fantastic figures of the Trois Frères Sorcerer; elaborate tools such as boomerangs, harpoons, and ropes; mineral quarries; ornate funeral rites; and many other wonders. As Randall White states, in *Natural History* magazine: "Cro-Magnons used two- and three-dimensional forms of representation systematically—to render concepts tangible, to communicate, and to explore social relations and technological possibilities. This powerfully enhanced their evolutionary fitness."[24]

Around this time we also see the rise of what must have been some form of organized religion, and belief in an afterlife, as can be gleaned from the 28,000-year-old site of Sungir, near the city of Vladimir, 100 miles east of Moscow. There archeologists have found three bodies (a man and two children) wrapped in painstakingly prepared ivory beads, arranged in dozens of strands. Each corpse was sheathed in thousands of such ornaments, and given that each must have taken about an hour to make, funeral preparations would have used up thousands of hours of work—per body.[25] These rites of mortification betray an exquisite vault in imagination and motivation compared with the simple cave burials with a stag's head at Qafzeh 100,000 years ago, or with the deer's jaw that commemorated the death of the Neanderthal child at Amud 60,000 years ago.

This vision, liberated perhaps by language and softened by a breaking down of intellectual boundaries within the human mind, was therefore the final flourish that took us from the cave of Qafzeh to the artwork of Lascaux to the space race, the atom smasher, and the gene probe. At least that is the theory, which sounds convincing, but does not explain why *Homo sapiens*, forged, most probably, in the ancient ancestral homeland of sub-Saharan Africa, developed the requisite loosening of neurones that took us to the moon. We may never find out the exact nature of these forces though we have much to thank, and to curse, them for.

Nor were these the only alterations that occurred to our brains as we slowly changed our shape and our conduct during our five-million-year walk (a slouch initially) from the trees of eastern Africa. They just happened to be the last ones. Along the way we picked up a host of other behavioral modifications, and while the final few alterations may have had the greatest impact on global status, we should not forget the others. As Irven DeVore, of Harvard University, puts it: "Anthropologists have always assumed that evolution carried the human species up to the dawn of modern society and then left us there. After that, it is presumed that culture took over as the shaper of our behaviour."[26] This is simply not the case. We should never lose track of the fact that we are still primate, hunter-gatherers, and that we thrived until very recently in a Stone Age world. We bear the scars of that hominid childhood, which we can see in both our bodies and our minds. They do not disfigure us, but they do influence our lives.

So let us move on to an examination of the bodily and mental effects of our evolution on ourselves and our planet. These are the lessons that we have derived from paleontology, and from the study of evolution. Whether we have learned them in time is a different matter.

The decisive point is that we are—to all extents—the same creatures who only embarked relatively recently on their African Exodus. And that has been a considerable influence on the way we act today. In *The Stone Age Present*, a study of how our hunter-gatherer past still influences our conduct today, William Allman writes:

> The rich tapestry of behaviours that make up our modern everyday lives—our choice of a mate, our ability to live together in large groups, our love of music and concept of beauty, our anger in reacting to infidelity, our occasional hostility toward people who look different from ourselves . . . all have deep-seated evolutionary roots that stretch back to the times when our ancient ancestors were struggling to meet the challenges of the world around them.[27]

In this way, evolutionary psychology has been used to explain a host of human tendencies, some realistically, others more fantastically. For example, our fears—of dogs, large animals, snakes, and dark places—are probably inbuilt, it is said, even though the real dangers in life today have more to do with roads and house accidents. "It doesn't take much imagination to realize that large, carnivorous animals, poisonous insects, and dark caves were fears shared by our ancient ancestors—and for a very good reason," says Allman. "Roadways and electric-sockets, on the other hand, have not been around long enough to make an evolutionary impression on our psychology."

Then there is that occasionally baffling question of why we pick a particular person for a spouse. In the past, theories have suggested that people search for partners who resemble archetypical images of their parent of the opposite sex (a notion put forward by Freud and dedicated to Oedipus). Others have argued that we seek mates with characteristics that are either complementary or similar to our own qualities. Neither interpretation is correct, says the University of Michigan's David Buss, one of the founders of evolutionary psychology. As he puts it: "Underneath we are driven by the same desires: men place a premium on appearance, while women seek providers of resources and status."[28] In general, a man tends to seek a woman who will be fertile and provide him with many offspring who, in turn, will pass on his genes to future generations. A woman desires the same goal though she faces a problem that does not affect a man. His investment in creating a pregnancy amounts to a contribution of sperm, an input that can take very little time and effort (as women constantly point out). But for a woman, after fertilization, there is a period of gestation that lasts nine months and which is usually followed by a term of breast-feeding, which in our hunter-gatherer past would have lasted several years. Weighed down by dependent children, women would have desperately needed the support of partners, as their tribe roamed through foraging territories or followed herds of reindeer or caribou. This requirement has resulted in a form of biological conditioning that governs our unconscious attitude to mate selection. A woman seeks generosity, maturity, and social status. The search is for someone who will invest time, energy, and assets when she is pregnant or breast-feeding and who will later contribute to the rearing and caring of her children as they grow up. Age is not particularly important because a man can father children in his forties, fifties, sixties, and beyond. What is sought is stability, a man who will stay and nourish a relationship, and also provide resources to keep the family secure and healthy. What is to be avoided is a partner who cuts and runs, leaving a woman holding the baby. It is a search for "dads not cads," as Allman puts it.

However, a woman's attractiveness, according to men, is rated according to her appearance—the more voluptuous the better. And the reason is fairly simple. Clear skin, full lips, good muscle tone—all signal one thing—that a woman is fertile and capable of bearing those children that a man instinctively desires. This pattern is affected by the society in which we live, but only slightly. In different cultures, ideals of physical beauty divide over the question of body fat. In societies where food is relatively scarce, plumpness is valued. In those where there is a general abundance of nutrition, there is a tendency to prefer thinness. (In the United States, the world's richest nation, we see a measure of this trend in the fact that the vital statistics of Miss America—that icon of female allure—indicate winners have become 30 percent thinner over the years.) Buss has investigated a total of thirty-seven cultures, from the Gujarati of India to the college students of America, from the people of Zambia to the citizens of Tokyo, and responses to his questions—What do you want from a man? and What do you seek in a woman?—have been uniform: men seek younger, physically attractive women, while women want partners who are mature and affluent. (These two "urges"—male and female—explain, say evolutionary psychologists, the fact that women tend to marry slightly older men, a propensity repeated in societies across the world, though with some interesting variations according to culture. In the U.S.A. both women and men state a preference for an age difference of about three years between spouses; in Colombia, it is five years; and in Zambia, seven years.)

Many modern mating rituals, such as one-night stands, dating agencies, and toy boys, disguise these trends, Buss admits. "But these are just short-term strategies to select candidates who might become long-term partners. From there, we seek other qualities." This last point is important. Our Stone Age drives only prepare a selection of mates for us. After this, personal preferences take a hand, usually directing us to mates who are similar in intelligence, personality, and attitude to ourselves. More to the point, we are also supposedly unaware we are making such selections.

Such arguments stress the importance of appreciating human evo-lution when trying to unravel our psychologies—an approach that has been ignored for too long. For that alone, evolutionary psychology deserves praise. On the other hand, its explanations can occasionally resemble convenient post-rationalizations that lack experimental proof. Our conduct is usually shaped by everyday problems—jobs, money, relationships—and these far more immediate concerns are what most rule our lives. Our evolutionary history merely provides a background, albeit an important one.

In any case, our dual approach toward partnership—young women, caring older men—could only have developed relatively recently, for it rests on the general observation that humans bond (nominally, at least) with only one partner. We do not have sex with all members of our tribes, unlike chimpanzees. And men—outside the seraglio—do not generally indulge in building up harems (as gorillas do). We pair for life, or at least a good few years. This was not always the case with our predecessors, however. As we have seen, there are clear indications that australopithecines displayed pro-nounced sexual dimorphism—bigger males, smaller females—so that the former could compete for access to packs of females. Clearly, we have moved away from this social structure (though our frames may still carry vestigial traces of those distant days, for men are still about 12 percent larger than women). If we had not changed, our social lives would today be governed by the presence of harems, "each dominated by one giant, middle-aged man, twice the weight of a woman, who monopolized sexual access to all the women in the group, and intimidated the other men," says Matt Ridley in *The Red Queen: Sex and the Evolution of Human Nature*.[29]

In addition, females evolved so that they no longer advertise the fact that they are ovulating and fertile. By contrast, chimpanzees do this very noticeably. Female bottoms swell and become very pink, attracting a great deal of male interest. The rest of the time, they get relatively little. So what kind of society would we live in if females advertised their fertility in quite such a striking way? "Sex would be an intermittent affair, indulged in to spectacular excess during the woman's estrus, but totally forgotten by her for years at a time when pregnant or rearing a young child," says Ridley. "Men would try to monopolize such females for weeks at a time, forcing them to go away on a 'consortship' with them, but would not always succeed and would quickly lose interest when the swelling went down."

As to the reasons for the evolution of concealed estrus, matters are far less clear. Indeed, the issue is dogged by controversy with a host of different ideas vying for academic credibility. It has been suggested this biological disguise developed so that men would not drop everything each time a female came on heat, diverting tribes from the important business of mammoth-spearing and mastodon-bashing; so that women would be attractive all the time and bind partners to them; so that men would be kept keen and busy supplying meat and other forms of nutrition to women; to confuse men about the paternity of children; and so that women would not use estrus as a means to avoid having children and so leave fewer descendants.

One of the most imaginative suggestions concerning this issue has been put forward by Dr. Chris Knight of East London University. He connects menstruation with the origins of culture, and in his book, *Blood Relations*,[30] Knight explores common elements in different hunter-gatherers' myths about menstrual blood and hunting. He believes these stories echo a cathartic event in human history, perhaps 100,000 years ago in Africa, when women brought about a social revolution—by going on a sex strike. In order to force men to organize collaborative hunting and share their food for the first time, women banded together to exploit their tendency to synchronize their monthly, menstrual cycles (an effect that can occur naturally, for unknown reasons, and which has been observed in large, cohabiting groups of women, for instance in army barracks). Female solidarity was enhanced by nonmenstruating women smearing themselves with blood or red ocher to mimic this lack of availability—hence the prevalence of the mineral at sites in Africa, Australia, and Europe. As a result, all the women seemed to be unavailable for sex, at the same time. So the male of our species, denied the joys of fornication, left home for the next best thing: hunting. Then, when these poor, deprived men duly returned carrying their antelopes and zebras, they were joined with the (now ovulating) females in a celebration of feasting, dancing, and sex. It may seem far fetched, but Knight's grand theory synthesizes many disparate anthropological and archeological observations and has attracted a surprising amount of serious attention.

Then there is the issue of pair-bonding. Why do men and women form such intense lasting relationships compared to the casual, temporary dallying that goes on with most other primates? Why do we mate, not necessarily for life, but often for very long periods? Some researchers suggest the tendency dates back to those days when *Homo erectus* women, having just evolved narrow hips to help them walk upright more effectively, faced the ancillary problem of having to give birth to neurologically immature children. Saddled with particularly helpless infants, we evolved a mutual support system in which males and females cooperated closely in the raising of their children.

However, some archeologists believe the tendency is a much more recent arrival. Dr. Olga Soffer, of the University of Illinois, maintains

53 Reconstruction of the burial of a Cro-Magnon man at Paviland, South Wales, 27,000 years ago. The burial was accompanied by red ocher and carved ivory.

that premodern humans lived in small groups with limited foraging territories and little social or sexual differentiation of tasks.[31] Individuals got by as best they could, and while mother-child bonds were strong, there were no wide family or kinship structures of the sort we find today. This idea has been taken further by Binford,[32] who has developed a striking and controversial theory about the Neanderthals, whose men and women, he says, lived predominantly separate lives, coming together only every few weeks for social or sexual exchanges. "It is an interaction we don't commonly see in modern humans," he states. "There is independent food preparation, different land use patterns, different uses of technology. In modern humans the relationships are more integrated. But the Neanderthals are separate, yet they are interacting."

Both Soffer and Binford argue that there was a major social transformation during the evolution of *Homo sapiens*, producing much greater complexity in social structure, based around extended nuclear families and wide nets of relatives. In this way were born dynasties, chiefdoms, and social class. Early evidence of this comes from the three 28,000-year-old Sungir burials where tremendous effort went into the production of the ivory beads which lay on the bodies—at least 2,000 hours of work in the case of the male corpse. He could easily have earned the great status bestowed him in death, of course. Yet the other two bodies were those of children who could not have had time before they died to achieve such rank. And in fact, more than double the effort was spent on their funeral preparations compared with the solitary male, which suggests an inherited status

54   An amusing comment on the debate about why the Neanderthals became extinct.

where their position was ascribed rather than achieved. And yet, there is no evidence of such intense, inherited social stratification in any premodern people.

It was perhaps the starting point of many of the heartaches that beset our modern civilization: elitism, status-seeking, wealth, and poverty. We shall explore these in a little more detail in the next chapter, along with an examination of how we carry around our hominid past, not just in brains, but in our bodies. It is a fascinating business, though ultimately not a comforting one.

9

Prometheus Unbound

I think it's one of the scars in our culture that we have too high an opinion of ourselves. We align ourselves with the angels instead of the higher primates.

Angela Carter

Broadly speaking we are in the middle of a race between human skill as a means, and human folly as an end.

Bertrand Russell

Take a glance in a mirror. Now look at your teeth: pearly white, or possibly yellowish? Don't worry, either way, their appearance makes little difference to their other attributes. They should have strength, durability, resilience, and most of all, diminutive dimensions—for unlike carnivores, who have large canines to rip flesh, or herbivores, who need powerful masticators to grind vegetation, we have petite teeth. But why? And while you are about this questioning self-perusal, look at your hair (if you have any). Why is it concentrated on your head? Why do men, but not women, have beards? And why do we have tufts on our armpits and groins?

You may well have asked yourself these questions over years of gazing wearily at your own increasingly sagging reflection. On the other hand, sheer familiarity may have dulled such interest—which would be a shame, for the looking glass reveals many telling clues about our evolution as intelligent, and ultimately world-conquering, primate foragers, and about our long march to "civilization."

Indeed, when we stand back and ask basic questions about our form, which Hamlet thought to be "so like a god," we can see a creature that, for all its twentieth-century grooming, has hardly changed from the hominid who made its African Exodus only 100,000 years ago. We should therefore not be fooled by our modern technologies and culture. Beneath that patina of urban sophistication—our haircuts, spectacles, makeup, and the rest—we still display attributes and physiques that have been honed on the African savannas over the past four million years. We come to our local environments carrying all these trappings. We are not biological devices newly constructed for the twentieth century. We are really just a "rather odd African ape,"[1] as Harvard anthropologist David Pilbeam puts it, a primate that has simply not had time to adapt significantly since our move from our birthplace. We still have Stone Age bodies, and therefore Stone Age brains, for what is true for our appearance must also be true for our behavior.

Clearly, staring at our reflected likenesses can be a provocative business and can prompt a welter of questions about the origins of our demeanor, intellect, and prospects. This book has explored the strange and dramatic story of our African origins, our journey out of the continent, our crossing of the rest of the world, and our final settlement in every conceivable nook and cranny on the planet's surface. Now it is time to look at the resulting stigmata, the features that we evolved as hominids and which now give us an awareness of who we are, and why we are in our present condition, anatomically and also culturally, politically, and scientifically. We are animals, albeit intelligent ones, and we should not think we are immune to the processes of evolution. So let us look at its handiwork, both ancient and recent, and its shaping of modern human beings. And let us begin with the first on our list: our appearance. We are primates, after all. So what can paleontology and anthropology reveal about the looks, and physiques, of *Homo sapiens*, and about the evolution of our physical features? The answer is: quite a lot. So let us commence our examination, starting from the top.

On our heads, we find hair that shields our brains from dangerous fluctuations in solar radiation, a fairly obvious conclusion that nevertheless begs a second, more pertinent question: Why do some men

lose that protection in middle age? Again the answer is straightforward. A gene (or perhaps genes) for baldness does not usually
manifest itself until after a man has passed his age for fathering children—at least in Stone Age terms when the average lifespan was less
than forty years. Losing this key protective covering would therefore
have made no difference to the furthering of his genotype. He would
have procreated by then, and so these "non-deleterious" genes have
maintained their hated hold on pates—as they have done throughout
the ages.

As for beards, it is generally believed these acted as visual devices
to enlarge men's faces, perhaps to frighten off enemies or impress
women. However, the distribution of "hairy faces" is uneven in both
hot and cold climates, and through differently pigmented races. "This
suggests once again that selection in favor of male beards was mainly
the result of prehistoric cultural preferences," says Jonathan Kingdon,
though tastes can change over time, as he also points out:

> Today sophisticated cultures seeking a sense of order and confor-
> mity in crisp, trim images are embarrassed by this reminder of
> prehistoric unruliness; the answer is to shave. Cultures that seek
> to appear civilised but also to maintain a firm distinction between
> men and women compromise—the men grow moustaches.

And those hair tufts on pudenda and armpit? They lie above
warm glandular areas and allow hormones and other scents to linger,
evaporate, and disperse their intimate messages. The flight of the
pheromone starts here. The rest is nakedness, the remainder of our
primate hair having thinned dramatically because our ancestors
evolved in hot, unshaded ground on the savanna plains of Africa. In
extreme heat, the body is capable of sweating off a maximum of
twenty-eight liters of water in a day. Thick hair would have reduced
the effects of that cooling and would also have allowed salt and other
wastes to build up in our fur or hair. Without that obstacle, sweat
simply washes off our bodies.

Moving down, we pass over the forehead, which is remarkably
prominent for a primate, the result of that relentless increase in the

volume, and change in shape, of *Homo sapiens'* crania, especially in the frontal lobes. Our brains have evolved narrower and taller dimensions, as if squeezed at the front, back, and sides, a physical metamorphosis that may reflect those neurological changes that gave us world dominance. In contrast our browridges are pygmy-sized by hominid standards, and are marked only by eyebrows, which may help keep sweat out of our eyes, though they may also perform a role as facial signalers that aid communication. On the other hand, our noses come in a variety of shapes and sizes, which, we have seen, are associated with climate: broad for tropical origin, beaky or narrow for colder zones, although other factors, including sexual selection, may also have been involved in the human hooter's evolution.

This facial peregrination leads next to the mouth and jaw, which provide perhaps the greatest interest: for they most clearly reveal technology's cutting role in the rise of *Homo sapiens*. Over the past two million years, our jaws have contracted as we have switched to a more easily digested diet of meat, offal, and marrow that is rich in protein, fats, and carbohydrates. We have also found ways to make these foods even easier to eat: knives to cut flesh, and fire to cook it, as well as grinders and pounders for pulping grain and seeds. This descent towards the diminutive began with *Homo erectus*, took a brief detour with the large front gnashers of Neanderthals, and returned, in earnest, with modern humans.

And the cause of this orthodontic decline is quite simple. With the provision of easily digestible food, our teeth no longer had to act as heavy duty food processors, and so, with the relentless economy of evolution, they have slowly shrunk as a result of this culinary emancipation. Our jaws have contracted for the same reason, and the primitive, muzzle-shaped skulls of our predecessors have been transmuted into flat-faced, modern heads, though with a heavy price attached— for the cues which regulate tooth growth are different from those controlling jaw development. The former is determined purely by our genes, which are assembled at conception, while the latter responds to both genetic and environmental factors. So even in people "programed" to have large jaws to hold large teeth, we find a reduction in the size of the former because they have been raised on white breads, pot noodles, pizzas, and all the other low-chew, take-out wonders of modern cuisine. As a result, people often end up with jaws that are too small to hold all their teeth. They may just about accommodate the first twenty-eight, but the arrival of the last four—the wisdom teeth, in early adulthood—is the final anatomical insult. A person simply has no room left in their mouth for them. So they grow in at an angle and get caught up with the other teeth, forcing them out of position; or they erupt out of the sides of the gums; or they get stuck inside gums, causing abscesses.

This is the price that we have paid for our progress in making tools and our discovery of fire—and the problem is widespread. In 1993, a quarter of all British people in their thirties had to endure at least one impacted wisdom tooth, of which 116,000 had to be removed. In a society blessed with plentiful antibiotics and dental surgeons, such sufferers survive. Elsewhere, they can die from serious infection—often in their teens—as they did in the West before such lifesavers arrived.[2] Such attrition creates evolutionary pressure, for those with fewer wisdom problems will survive longer to have more children. And this is exactly what we see. In some countries, people are no longer being born with all their wisdoms. Some have only twenty-eight teeth—as opposed to *Homo sapiens'* normal full set of thirty-two. Among Europeans, for example, it has been found that up to 15 percent of people have at least two wisdom teeth missing (i.e., ones that never grow in at all) while in east Asia, the figure can be as much as 30 percent in some areas. It is a classic example of evolution in action—except, of course, that in the West the trend has already been halted by modern medicine. Thanks to its intervention, having thirty-two teeth is no longer a threat to one's prospects of procreation. So, the relentless hand of evolution has been arrested—by dentists.

Finally we come to the chin, another special human stigmata. It encases and strengthens the jaw—from the outside. In our ancestors, such as *Homo erectus*, this reinforcing took place inside. We had to develop an external supporting casing because our jaws were getting smaller, and our tongues needed plenty of room to move about, especially as our blossoming linguistic powers demanded a greater flexibility of sound production. In short, we have a "talker's chin."

Of course, the very fact that we are most likely to carry out this facial scrutiny on a wall-mounted mirror is a reflection of the most important of all our evolutionary developments over the past five million years—our adoption of an upright stance and gait. As we have seen, it left hands free for tool design when the human brain eventually began its enlargement and onward push to consciousness and intelligence. "Upright posture is the surprise, the difficult event, the rapid and fundamental reconstruction of our anatomy," writes Stephen Jay Gould in his essay *Our Greatest Evolutionary Step*:[3]

> The subsequent enlargement of our brain is, in anatomical terms, a secondary epiphenomenon, an easy transformation embedded in a general pattern of human evolution. As a pure problem in architectural reconstruction, upright posture is far-reaching and fundamental, an enlarged brain superficial and secondary. But the effect of our larger brain has far outstripped the relative ease of its construction.

We sadly take our two-legged prowess for granted, says Gould. "It is now two in the morning and I'm finished," he concludes. "I think I'll walk over to the refrigerator and get a beer; then I'll go to sleep. Culture-bound creature that I am, the dream I will have in an hour or so when I'm supine astounds me ever so much more than the stroll I will now perform perpendicular to the floor."

The move to upright gait involved major anatomical changes all right, but not all were beneficial. Walking on two legs produces greater wear and tear on hips, which have to bear our entire body weight. In other primates, this load is shared over four limbs. Bipedal humans pay the price through disablement, in our later years, that can only be put right through hip replacement operations. Such medical intervention is also a consequence of our increasing longevity, of course. However, other side effects intrude at a far younger age, the most profound emerging because we have evolved relatively narrow hips and pelvises. Had we retained the wide hips of apes, our centers of gravity would have swung around all over the place when we walked, and our gait would have become enormously draining in terms of energy expenditure. So we adopted a narrow, cylindrical form and stance, with knees tucked underneath our bodies. (We may take pride because *Homo sapiens* is "the smart ape," yet we could equally call ourselves "the knock-kneed primate.") This posture has allowed us to stride across the globe—but at a cost.

Our hips form a bony ring—the pelvis—and possessing narrow hips produces a tight pelvis. Given that a child must pass through this constricted gap when it is born, this meant that human anatomy faced a drastic challenge once our cortexes began to expand. "No rational animal would knowingly be bipedal and have large brains," says Leslie Aiello, of University College London.[4] "The consequences for women have been horrendous." For a start, there is the issue of brain growth. In other primates, indeed in nearly all other animals, this stops after birth. An ape goes through its most crucial neurological development within the womb. A human being is born having undergone less than half this critical increase, for the simple reason that our heads would be too big to pass through our mothers' pelvises if our cortexes were allowed to reach their full size. "A human pregnancy would have to be about twenty-one months long if we were to be born as fully-fledged neurological entities," adds Dr. Aiello:

> We are not, of course, and so a child has to spend the first year of its life in a particularly helpless form as it completes the brain development that should have occurred while it was a foetus. This in turn constrains the mother. She is tied to an utterly dependent young child for a long period, yet she needs good nutrition to provide milk rich in protein, fats and carbohydrate for her offspring. She is therefore particularly reliant on the support of her spouse and the rest of her group.

This process probably began with tall, slim *Homo erectus*, which was capable of striding across the savanna with an ease that dwarfed that of its australopithecine and *habilis* predecessors, and which also started to evolve significantly larger brains. The consequence was that neurologically immature infants were born for the first time, a trend that has become more pronounced as the millennia have passed. Yet even with the delivery of babies in extremely early stages of development, which occurs today, problems still occur. Other primate babies can pass straight through the pelvis. A human child has to twist through the narrowest of gaps like a cork being pulled from a wine bottle—a maneuver that requires much effort from the mother, and the attention of midwives. Even with the intervention of modern medicine, however, human birth remains a surprisingly risky business. This conflict of function also takes its toll on women, who are slightly less efficient bipedalists than men because of their roles as potential mothers. Women have wider hips, to reduce the trauma of childbirth as much as possible, and pay the price by being slightly poorer walkers, runners, and jumpers, as is demonstrated by their somewhat inferior Olympic running and jumping records. (Women's higher fat levels, and lower muscle mass, also play a role in limiting their ability in many sports, however, while enhancing their ability at long-distance swimming.)

Yet, men should not let their prowess run to their heads, for it is equally clear that, over the past 10,000 years, their athletic performances have probably deteriorated, despite the fact that records are continually broken at major competitive events. The reason for this decline provides the story of our evolution with one of its most bizarre footnotes—for, quite simply, *Homo sapiens* is shrinking. Research from several continents reveals that the human race has been waning anatomically for the past ten millennia. Men today are about 5 ft. 8 in. to 5 ft. 9 in. tall in the West. Cro-Magnon males were about 6 ft. Even our brains have been getting smaller, as measured by our skulls, which are 10 percent smaller than those of Cro-Magnons. (And think what this discovery does for the ideas of Rushton and the other racial evangelists. By linking cranium volume so directly to the concept of intelligence, they are left struggling to find an explanation why, according to their ideas, the human race must be getting stupider.) Of course, our cranial and anatomical decline is not a major business. Our species did not begin like Stone

Age Schwarzeneggers and is not likely to end up dwindling to the status of vertically-challenged Woody Allens. Indeed, the shrinking may already have stopped. Nevertheless, the effect is real.

This discovery may seem surprising given that so many signs point to recent size increases: museums with tiny suits of armor, old houses with low doorways, and diminutive four-poster beds. In fact, people were smaller in medieval times because they ate so poorly, and so did not reach their full, genetically programed size. That is why we seem to have grown since then. The effect is just a blip, however. The real underlying trend is one of decreasing stature.[5]

But why? What possible advantage could there be in having reduced dimensions? One suggestion put forward by researchers is that more effective hunting techniques (traps, spear-throwers, and bows and arrows) meant we no longer required the strength and size demanded by life in earlier millennia. This change was accentuated by farming, which began about 10,000 years ago, and which freed men and women from the rigors of hunting, and searching for roots and berries. It lessened the need for robust physiques and we have been shrinking ever since. Unfortunately, this explanation is confounded by Australian data. "Agriculture did not arrive here until Europeans first stepped ashore 200 years ago," points out Peter Brown, of the University of New England, whose uncompromising views on human evolution in Australia were revealed in Chapter 6:

> Native Australians were hunter-gatherers until then. Yet exactly the same ancient decline in stature can be seen in their fossils. Ten thousand years ago, aborigine men were between 5 ft. 9 in. and 6 ft. tall. Today, they are between 5 ft. 5 in. and 5 ft. 6 in. More to the point, humans were not the only species which shrank. In Australia, every animal bigger than a wombat got smaller, and if you don't know how big a wombat is, try to imagine a corgi dog on steroids.[6]

It is a trend observed by other researchers who have noted that in Europe, both hunter and prey have been shrinking over the past ten millennia. And if every animal bigger than a "pumped-up corgi" became smaller, then a more general phenomenon than the mere rise of agriculture must be involved, adds Brown. "Climatic change was probably the key. The world began to get warmer 10,000 years ago as the last Ice Age ended, and that must somehow have triggered our size reduction."

However, a far more bizarre explanation is put forward by Professor Majie Henneberg, of Adelaide University. His research has not only demonstrated that a "significant decrease in cranial capacity"[7] occurred in *Homo sapiens* over the past 10,000 years, but has also revealed that birth weight—which is related to stature—continues to vary to this day. After studying birth records at South African hospitals, he found that children born between May and October are 11 percent lighter than those born the rest of the year. Bizarrely, Henneberg has discovered similar fluctuations in the birth weights of dogs. "This is probably related to the position of our planet on its elliptical orbit," says Professor Henneberg. Gravity and electromagnetism vary, and probably produce those annual changes in human and dog size, as well as in other animals, he argues. And if minor stature-changing fluctuations in radiation and growth occur round the year, then there may be far bigger alterations taking place over several millennia, ones that may be responsible for those drops in human size.

Unsurprisingly, not every scientist agrees with this radical interpretation. Most researchers point out that even in places where agriculture was never invented, such as Australia, there were clear technological improvements in stone toolmaking and abilities to hunt game. Population densities went up about 6,000 years ago in many such places, across the world. This trend would have increased until it reached a limit, when food would have become scarce again. Humans therefore had to drop in either number or size, and evolved the latter course. (However, such a theory does not explain the drop in size of other animals—unless competition with growing numbers of humans acted as the trigger.)

Regardless of which explanation one chooses, it is clear that around 10,000 years ago human food-gathering behavior began to change profoundly, with the birth of agriculture transforming hominid foragers into primitive farmers. With the availability of a reliable food supply nearby, the first sedentary societies were created, with their villages, towns, and eventually their cities and civilizations. The concept of individual land ownership, which most of us now take for granted, came into being, and humans were turned into new, highly intensive producers of food. This manipulation of plants and animals started, independently and within a relatively short time, in several areas of the world. Wheat was cultivated in the Middle East, rice in China, maize in South America, and sorghum, millet, and yams in West Africa. A major driving force in this change was the alteration that was then going on in the climate. The last Ice Age had ended, and sea levels were rising. The weather became hotter and moister, and in areas where foragers kept to a timetable of well-defined visits to various sites centered upon food sources, they would have noticed that the seeds and plants they discarded had begun to grow on their return. "The concept of possessing plants and the places they grew in, various interventions to promote growth and possibly some weeding must all have preceded horticulture," says Jonathan Kingdon:

> What was lacking was the idea of systematic planting and tending. That idea must have arisen in areas where growing conditions were very favourable, where human densities were high and territories were small. Also, I think it is a fair bet that horticulture was begun by women because they were more tied to the home base by their children and were better placed to tend plants while the men were away trapping, fishing or hunting.[8]

But if females were responsible for the birth of agriculture, they have paid a high price for their achievement. Researchers such as Theya Molleson, of London's Natural History Museum, have traced the effects of agriculture's development on the bones of ancient people, such as those from Abu Hureyra, a Neolithic (New Stone Age) settlement in Syria. There, it has become clear, the grind of everyday life quite literally marked the anatomies of the world's first farmers and, to a far greater extent, their wives. "Here, beside the

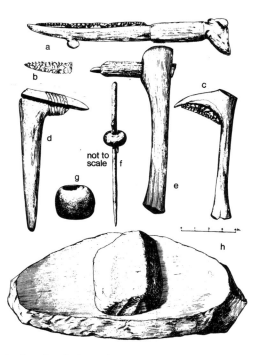

55   Simple agricultural tools from the Levant (a), Egypt (b), Europe (c, d, e, h,),
and Africa (f, g).

Euphrates, *Homo sapiens* was transformed from a hunter-gatherer
into a farmer," says Molleson.[9] "At the same time, women were tied
to their place of work for the first time."

It is at places like Abu Hureyra that we see the first widespread
appearance of those perennial ailments of life today—back problems,
aching limbs, and much more. Molleson has studied the bones of 162
individuals who lived at Abu Hureyra from about 11,500 to 7,500 years
ago, and found injuries indicative of demanding physical activity—but
only among farming people, not among their hunter-gatherer predecessors. There was vertebrae damage, severe osteoarthritis in toes,
curved and buttressed femurs, and knees with bony extensions. At first,
Molleson blamed sport or dance. "But crippled ballerinas seemed
unlikely during the Neolithic period," she adds. Then the true cause
became apparent. "With the advent of agriculture, men cultivated
plants, while women took on the job of grain preparation." Women had
to kneel before saddle querns—plinths on which they rolled stones to

crush corn. "Kneeling for many hours strains the toes and knees, whereas grinding puts additional pressure on hips and the lower back," Molleson writes. The result: damaged discs and crushed vertebrae.

In addition, bones rubbed on bones, damaging cartilages, while arthritis affected toes that were constantly pressed down to provide leverage. These were the repetitive strain injuries of the Stone Age and most were found on female skeletons. These endeavors made food supply more certain, but also generated other problems.

Coarsely-ground grain, with lumps of kernel remaining, had "an appalling effect on everyone's teeth." The solution was the next technological development: the sieve. With the advent of weaving, women were able to make containers in which to sort the chaff from the wheat, and so eliminate those kernel-cracked canines. But to do that, women had to hold canes or straws in their mouths as they manipulated the other strands of a basket or sieve. The result: grooved teeth, which were also revealed in Molleson's studies.

Next in this technological progression came the creation of pottery—about 7,500 years ago—that allowed women to soak and cook grain. "One result was porridge, which soon had a dramatic effect on society," adds Molleson. Nutritious gruels could be given instead of breast milk, while mothers could eat carbohydrate-rich foods. The result: early weaning and increased fertility. (It is worth noting why this would have happened. Regular breast-feeding releases hormones that suppress ovulation, a phenomenon that has probably evolved so that a hunter-gatherer mother would not face the problem of carrying several babies, each requiring milk. But by taking children off the breast at an early age, this hormone signal is disrupted and ovulation restarts earlier than nature anticipated. The result is smaller gaps between pregnancies. The baby boom began here, in short.)

Agriculture had a devastating effect on health. Although it is often promoted as the great technological achievement that freed men and women from the drudgery of hunting and gathering food, farming took a grim toll on its practitioners. In another study, this time of Native American skeletons excavated from burial mounds in Illinois and Ohio, scientists have found evidence of an even bleaker physiological transformation, one that occurred with the arrival of the cultivation of maize (i.e., sweet corn) about 1,000 years ago. Healthy hunter-gatherers were suddenly turned into sickly farmers. Tooth cavities jumped sevenfold; children's teeth defects reveal their mothers were badly undernourished; anemia quadrupled in frequency; tuberculosis, yaws, osteoarthritis, and syphilis afflicted large numbers of the population; and mortality rates jumped. Almost a fifth of the population died in infancy. Far from being one of the blessings of the New World, corn was a public health disaster.[10]

It is a story repeated time and again round the world. With the arrival of farming, populations soared while the health of individuals plummeted. Nor should we be that surprised when we look at the statistics. Bushmen in the Kalahari Desert base their diet on eighty-five different wild plants. They are an exception, however. Most of the rest of the world is fed by farmers. As a result, just three plants high in carbohydrate—wheat, rice, and maize (corn)—provide more than 50 percent of the calories consumed by the human race today. We get our calories cheap, in exchange for poor nutrition. And sometimes this dependence can go catastrophically wrong, as it did in the 1840s, for example, when a blight attacked the potato, the staple crop of Ireland, leaving more than one million people to starve to death. For a hunter-gatherer, who finds food all around, such an idea would seem inconceivable. In addition, the establishment of dense populations of people, who could exist on grain and rice stores, triggered many of the world's deadliest epidemics, diseases that thrived among cramped, underfed peoples. Tuberculosis, leprosy, cholera, and malaria all appeared in the wake of farming. Similarly smallpox, the plague, and measles only manifested themselves with the arrival of cities.

Now this catalogue of woes may sound odd to Western ears. The idea that human beings could be better off as hunter-gatherers seems ridiculous. However, as Jared Diamond points out:

Americans and Europeans are an elite in today's world, dependent on oil and other materials imported from countries with large populations and much lower health standards. If you could choose between being a middle-class American, a Bushman hunter, and peasant farmer in Ethiopia, the first choice would undoubtedly be the healthiest one, but the third choice might be the least healthy.

A recognition of this point has led to the rise of an intriguing new source of health remedies: Stone Age—or Darwinian—medicine. Proponents of this fledgling science point out that although antibiotics now keep people alive, often until their eighties, we are not necessarily that healthy. As Randolph Nesse and George Williams, two of the discipline's founders, argue: "Advances in medicine would be even more rapid if medical professionals were as attuned to Darwin as they have been to Pasteur."[11] Take the simple question of diet. *Homo sapiens*, as hunter-gatherers, clearly lived on a wide variety of vegetables and free-range meats. Dairy foods—butter, milk, and cheese—played no part in our eating habits for most of our history, for they only arrived on our tables in the wake of the Agricultural Revolution. These are "unnatural" products, high in fats and other cholesterol-raising constituents, and we should avoid them. It is a point reinforced by Boyd Eaton of Emory University, Atlanta. As he puts it: "Modern humans are Stone Agers displaced through time." We are probably programed to store fat reserves against possible lean times, for example. In the West, these no longer exist, but we still keep storing away. The result: overweight Westerners. Equally, we may be conditioned to minimize physical activity when it is not absolutely necessary—an adaptation to conserve food stores. So we loll about in front of our televisions. In short, we can blame the fact we are fat couch potatoes on our Stone Age ancestry.

The answer is take exercise and eat non-fatty foods, which is scarcely a revolutionary idea, of course. On the other hand, an awareness of why such a lifestyle is important—gleaned through an understanding of what kind of creatures we really are—helps reinforce the point, and may help people be more healthy. In any case, scientists such as Eaton have other ideas. For example, they point out that women in the West face greatly increased levels of various cancer such as those of the breast and ovary. Part of the problem lies with the fact that girls generally reach puberty earlier, but have children later, than they would if hunter-gatherers. Our lifestyle seems to induce cancers, and if we could hormonally re-create the chemical balances that must have existed in Stone Age times through courses of injections, we might be able to reduce cancer levels in women, say the Darwinian doctors.

The discussion of our sickly lifestyles raises a simple question, however. Why on earth did *Homo sapiens* turn its spears into ploughshares in the first place? If agriculture has brought such medical disasters down upon our heads, why did we plump for lives tilling the land? The answer is simple: sheer force of numbers. Farming a piece of land can support ten times the number of people that would live off it as hunter-gatherers. As a result, relatively healthy Bushmen, and others, have been marginalized into some of the planet's worst real estate by the masses of poorly nourished farm workers who are still spreading across the globe. It is one of the most important lessons that archeology can teach us. Given a choice between limiting our numbers or trying to find more food to eat, we have consistently plumped for the latter course of action: despite the fact that it has brought us pestilence, starvation, and war. Yet signs that *Homo sapiens* possesses the ability to learn, eventually, the significance of our agricultural past remain fairly dispiriting. By the end of the millennium, there will be more than six billion inhabitants on planet Earth, and a great many of them will be more poorly nourished than were their Stone Age predecessors of 20,000 years ago.

The effect of rocketing numbers of human beings upon Mother Earth has been grotesque, both in terms of our active and passive participation in zoological slaughter and ecological devastation. Just consider humanity's impact on the fauna of North America. As we saw in Chapter 6, the Clovis people eleven millennia ago appear to have had a fairly devastating effect on the continent's wildlife, wiping out an estimated seventy-five species of large creatures such as the mammoth—though that was nothing compared to the mayhem of recent years. After their initial bout of slaughter, native hunter-gatherers achieved a relatively stable coexistence with the remaining animals of the continent. The bison was a principal resource of the

Plains Indians, for example, furnishing them with food, skins for shelter and boats, bones for tools and utensils, and "buffalo chips" (dung) for fuel. Then came white men, and their guns. From the grasslands of the Mississippi River to the Rocky Mountains, thirty million prairie bison were wiped out until their numbers were reduced to about 500 near the end of the last century. (An estimated 35,000 to 50,000 now live on refuges and ranches today.)

The bison was not the only species hounded to near-extinction, however. In *The Endangered Kingdom* Roger Di Silvestro recounts how on a single eighteenth-century hunt in Pennsylvania, hunters from across the state formed a circle 100 miles in diameter and marched inward, blasting all that they encountered. A total of 41 cougars, 109 wolves, 18 bears, 111 bison, 112 foxes, 114 bobcats, 98 deer, and more than 500 smaller mammals were killed in the process.[12]

Similarly, the arrival of the first people on the South Pacific islands led to the extinction of 2,000 bird species alone—through hunting, egg collecting, habitat disturbance, and more recently with the spread of other species such as pigs, dogs, and rats. It is a bloodbath that has been repeated countless times across the globe. First, Stone Age hunters eradicated large creatures unused to human predators, then a period of equilibrium was established, before a second wave of westernized settlers arrived to begin the bloodshed again with a fervor and with an arsenal of unprecedented effectiveness. As Peter Ward points out in *The End of Evolution*: "If mankind could so quickly destroy the majority of the world's big game with a primitive Stone Age technology, what hope have the world's creatures in the face of our far more advanced technology?"

Documented dispassionately, it is clear that ours is the behavior of a species no longer living in harmony, or more prosaically, in balance with nature. The consumer and consumed are out of kilter with each other. Just consider that rise in human numbers that we mentioned earlier. We may now be approaching the birth of the six billionth member of our species. Yet *Homo sapiens* only reached its first billion about A.D. 1800. By 1930, it had achieved its second; its third in 1960; fourth in 1975; fifth in 1987; and, by 1992, 5.5 billion—with a large proportion living in abject poverty. The most optimistic scenario suggests the world's population will peak at around ten billion people in the year 2100, at its worst about fifteen billion. And this is not some species of beetle, that exists on scraps of foliage, that we are discussing. This is *Homo sapiens*, an omnivore fresh out of Africa that requires 3,000 calories of energy a day but which has nevertheless soared from around ten thousand individuals 100,000 years ago to an estimated ten billion in the near future—a millionfold increase in a few millennia. "Our planet cannot withstand such numbers," says Peter Ward:

> To realise enough organic productivity, virtually the entire arable land surface of the earth—every forest, every valley, every bit of land surface capable of sustaining plant life, as well as much of the plankton of the sea—will have to be turned over to crops if our species is to avert unprecedented global famine. In such a world, animals and plants not directly necessary for our existence will probably be a luxury not affordable. Those creatures that can survive in the vast fields and orchards will survive. Those that need virgin forest, or undisturbed habitat of any sort, will not.

In short, our chances of maintaining our few wildlife refuges—currently our only hope of staving off most major bouts of extinctions—are negligible. How can we expect people to tolerate the presence of edible, or saleable animals, when they or their children are starving? After all, would New Yorkers and Londoners be able to keep their hands off a refuge in the Bronx or Regent's Park if they were suffering severe malnutrition? The fact that we expect Africans or Asians to show such restraint reveals our inability to envisage the enormity of our sphere of influence today. We are prisoners of a limited mentality, a handicap that prevents us from truly understanding numbers of humans greater than a few hundred. Our lack of large-scale empathy may soon kill us, however. Apart from taking over every ecological crevice on the planet, we are also poisoning its air, land, and sea. Every year, five billion tons of carbon are burned as fossil fuel, increasing carbon dioxide levels by almost 5 percent a decade, a rise that has taken the gas's concentration of 275 parts per million (ppm) in the year 1850 to a level of 345 ppm in 1985 and beyond. We are destroying the rain forests of the Amazon and West Africa, removing the very trees that absorb that carbon dioxide and return it to the air as oxygen which we can breathe. Friends of the Earth estimates that we are currently destroying rain forests at the rate of six football fields a minute.[13] At this pace, in the next forty years all the world's rain forests will have been wiped out. Such an atmospheric insult threatens to trigger a calamitous rise in temperatures round the globe, eventually melting ice caps, flooding coastal zones and islands, and producing "a refugee problem of unprecedented proportions," in the words of Robert Buddemeier, an environmental scientist at the Lawrence Livermore Laboratory, in California.[14] Adding to this there is the fact that in the West, about twenty-five tons of rubbish are produced per head of population per year. Not all is household junk either. Most is industrial, often toxic, waste. In addition, seas are polluted, their fish stocks are now dangerously depleted, and even the rain is turning acidic, damaging forests and poisoning lakes. Meanwhile, over the northern and southern hemispheres holes in the atmosphere's ozone layer, which protects us from the sun's harmful ultraviolet rays, are appearing because we are pumping industrially produced chemicals into the air. This litany of environmental grief is not a new one, of course, and has been somewhat oversold in its constant retelling. We have become inured to its message, which is a shame, for the warning is alarming, and becoming increasingly urgent.

Nor are our ecological and overpopulation woes the only problems that now afflict us. As our technology becomes more and more sophisticated and speedy we face the prospect of simply being overwhelmed by it. Take this quote from an operative inside the control room of the Three Mile Island nuclear plant, near Harrisburg, when it came perilously close to a devastating meltdown: "Bells were ringing, lights were flashing, and everyone was grabbing and scratching." In other words, events were happening so speedily, and data was spewing out so quickly, that humans just could not respond quickly enough. As Erich Harth states in *Dawn of a Millennium*: "Human beings are not wired for such speeds and such informational deluge."[15] The shooting down of the Korean Airlines Flight 007 in 1983, the destruction—by a missile launched by the U.S. cruiser *Vincennes*—of an Iranian passenger jet in 1988, and the explosion which wrecked the Chernobyl nuclear reactor in 1986 are other examples of human beings overwhelmed by their own sophisticated machinery, adds Harth:

> Technology is cumulative, growing through the addition of many small contributions, while intelligence, the source of this steady growth, remains fixed. At a certain point, we may find ourselves overwhelmed by our creations, when the intelligence required to achieve a certain level of technology may be less than that needed to survive it.

The most worrying of these technological avalanches, triggered inadvertently by hapless humans, is that of global nuclear war, a threat that has receded since the demise of the Soviet Union. In its place, however, we face the prospect of its stocks of plutonium and uranium being smuggled to terrorists across the world. Our technology—which took us to new continents and world dominion— threatens to rebound on us. As we have seen, the neuronal or behavioral differences between us and Neanderthals, which promoted the former at the expense of the latter, must have been very slight indeed. We may simply have stopped short of the full changes needed to control our own Promethean creations.

Of course, it may be that no species or planet can withstand the consequences of an accelerating technology once it has been set in motion: its strain on minds and environments may simply be too great. Unfortunately, we have no other examples with which to make comparisons. The last creatures, other than *Homo sapiens*, known to possess a recognizable technology disappeared from this planet 30,000 years ago when the Neanderthals perished, huddled in caves like Zafarraya. Nowhere else on earth, or in the universe, have we detected signs of anything comparable to the high-tech, electronic culture that now envelops our planet. This failure may just be bad luck, or it may reflect the fact that it is the fate of all creatures to succumb to their own technology. Certainly, decades of effort in the business of SETI (the Search for Extraterrestrial Intelligence) have produced no successes. As the great physicist Enrico Fermi once asked about alien civilizations: "Where are they?"

We have no answer to this question. However, we can at least respond to a more realistic one: What are our chances of encountering other technological cultures elsewhere in the galaxy. The American astronomer Frank Drake attempted to assess this probability by expressing the number of advanced civilizations that might exist in our galaxy. This calculation depends on seven terms including the chances of a planet having conditions favorable to life, and the probability of life evolving given these conditions. The last variable— the likely length expectancy of a technological civilization—was the most critical, however. If this figure is high, say millions of years, then Drake's equations produce a fairly large number for alien intelligences in our galaxy. Our interstellar neighborhood would therefore be teeming with travelers as populous and varied as those dubious denizens of the *Star Wars* bar. The opposite appears to be the case, however. Despite the claims of committed ufologists and flying saucer spotters, our skies have been remarkably clear of alien rockets and the like. We should take note.

Evolution is an inconstant business, of course, as Steve Jones observes.[16] "How could it have been predicted only 30,000 years ago that one moderately common primate would be among the most abundant mammals while its genetically almost indistinguishable relative was near extinction?" Is it possible that we may continue to change, to alter our behaviors, cultures, and brains even further until we have passed through our technological danger zone? According to Jones, the answer is almost definitely no. For *Homo sapiens*, the end of the line has been reached. The great African events that molded a creature of stunning adaptability have been halted—by our own handiwork, our culture. "In the West, most babies born now survive until they themselves have babies, so that existence is less of a struggle than it was," says Jones. "Nautal selection involves inherited differences in the chance of surviving that struggle, and as most of us do survive nowadays until we have passed on our genes, the strength of selection has decreased." The *Homo sapiens* of tomorrow will not have X-ray eyes or computers for brains because we have created a technology that does that for us.

However, that does not mean we have moved beyond the grasp of natural selection completely, as Christopher Wills has argued.[17] "Consider the evolutionary consequences of the shrinking ozone layer that protects Earth from shortwave ultraviolet light," he points out. Because of its emergence, people will have to wear hats and sun creams more often in the near future. However, he notes:

> Hundreds of millions of people on the planet have no access to sun block and must depend for protection on the melanin pigment in their skin. Of course, the amount of melanin in people's skin varies, leaving some far more vulnerable to ultraviolet harm than others. Those with pale, freckled complexions are already 4 to 20 times more likely to develop melanomas, a risk bound to increase as the ultraviolet flux increases. Unless we patch the ozone shield, this new kind of selection could weed out the fairest of us all.

In short, the fickle hand of evolution is unlikely to leave us untouched, though the forces that shaped this African interloper will make their impact on a far wider canvas than those that transformed us so radically 100,000 years ago. These were confined to one continent. That is not true today. We may all be Africans under our skin, but we are all global villagers as well.

Of course, the recognition that all human inhabitants of our planet share a recent African birth is startling. Indeed, this realization is so new that its significance is still sinking in, but we will never be able to look at ourselves in the mirror in the same light again. Just consider the political implications. Already the story of our African Exodus has entered the maelstrom of American racial issues. We have seen how Rushton has used our theory to account for the claimed primitive intellects and behavior of Africans, and how his views have been used in turn to "explain" the IQ inferiority of blacks, reported by him and the authors of *The Bell Curve*. They now provide ammunition for the American Right in their fight against welfare and affirmative action programs. But moderates in the U.S. can argue equally strongly that the close relationships between races implied by the Out of Africa theory suggest that intellectual differences can only be slight, or are mainly due to environmental and social factors.

And, not surprisingly, our theory has made a great impact in black communities, particularly in the U.S. Pressure from organizations such as the Tu-Wa-Moja African study group recently led the Museum of Natural History in Washington, D.C., to close parts of its exhibition on human evolution because it did not reflect Out of Africa thinking.[18] Part of their critique centered on the anthropological writings of the late Senegalese scientist Cheikh Anta Diop, who argued that the first Cro-Magnons closely resembled present-day African populations, and carried their ideas and art across the Straits of Gibraltar. The issue has even become entangled in the mythology of black separatist groups such as the Nation of Islam, whose leader Elijah Muhammad taught that all humans were black until an evil scientist, Yacub, produced a "bleached-out" white race through genetic experimentation thousands of years ago. Other Afrocentrists and black supremacists have turned the eighteenth-century European idea of nonwhite degeneracy on its head to argue that, by losing their melanin skin pigment when they left Africa, whites became inferior to blacks. "Melanists" argue that because blacks have higher levels of this substance in their bodies they are more sensitive and coordinated than whites.[19] However, there is no good scientific evidence to support this view either.

In fact, the really exciting issue concerns not the interpretation of old data, but the prospects for acquiring new information. So what new scientific data can we expect as we look forward to the next millennium? What surprises might we anticipate and how might they be discovered? Well, for a start, a new species of *Australopithecus* has been found at Kanapoi and Allia Bay in Kenya, and dated at about four million years old. Reports suggest it has arm and leg bones of a rather human form with jaws and teeth which look more like those of the old Miocene apes.[20] These Kenyan finds promise to further complicate the early hominid story. Similarly, studies of australopithecine sites in South Africa are providing further evidence of the chimplike anatomies and lifestyles of Dart's metaphorical children. In particular, their environments, it now seems, were wooded, suggesting that even about three million years ago, our predecessors were still fairly tree bound.[21]

Moving on through time and space, the Atapuerca sites continue to produce fragmentary finds of the first Europeans (more than 700,000 years old) who already show important differences from *Homo erectus*.[22] In addition, an astonishing find from the Altamura sinkhole in Italy awaits study. An early Neanderthal appears to have fallen into it, and starved to death, leaving his or her complete skeleton to be encased in stalagmite.[23] Other reports include the discovery of a nearly complete Neanderthal infant's skeleton in Syria;[24] a 100,000-year-old early modern skeleton from Egypt; the poignant 25,000-year-old burial in an Italian cave of a Cro-Magnon woman who apparently died in childbirth, her hip bones still enclosing the skeleton of her unborn baby.[25]

Similarly, new techniques are ferreting out the secrets of existing relics. Computed tomography (CT) scanners (used in large hospitals) can take remarkable three-dimensional X rays of fossils.[26] For example, the Boxgrove tibia is being compared with *erectus* and Neanderthal shinbones this way. In addition, skulls still embedded in rock can be scanned and virtual 3-D images, or solid replicas, can be produced by a laser beam which sets liquid plastic (a technique called stereolithography). This technique has produced the most complete image of a young Neanderthal child yet seen. Five fragments of its skull were found at Devil's Tower in Gibraltar, in 1926. However, it was only possible to fit two of them together, raising the prospect they may have come from two different children. But in 1995, a Zurich team used CT scans to fill in the missing parts—and proved they really did all belong together. The technique also provided unprecedented accuracy in reconstructing the size and shape of the child's brain cavity, and its skull thickness. Moreover, normally hidden structures, such as unerupted teeth, sinuses, and ear bones could all be visualized and re-created, revealing the child had suffered a jaw injury that disturbed its tooth development, and that its hidden anatomy was as distinctive from ours as its surface features.[27]

On a finer scale, scientists are probing fossils down to atomic levels. Scanning electron microscope images have revealed where the Boxgrove tibia had probably been gnawed by a wolf, and, by counting the weekly growth lines on its teeth, have also confirmed the age at death (about four years) of the Devil's Tower child. These techniques are also showing that the australopithecines, *Homo habilis* and even *Homo erectus*, had shorter childhoods than us. And by measuring trace elements and isotopes absorbed by their bones and teeth, their diets can be reconstructed. Results suggest the Neanderthals were indeed highly carnivorous, but that robust australopithecines (supposedly vegetarian) were apparently more omnivorous than we thought.[28]

Most tantalizing of all, however, is the prospect that ancient DNA will be extracted from extinct hominids, throwing a direct light on their relationship to us.[29] The breakthrough in molecular technology that allows tiny fragments of DNA to be located and copied in vast quantities—the polymerase chain reaction (PCR)—was made by the biochemist Kary Mullis in 1983, who was awarded the Nobel Prize for his discovery. As a result, scraps of ancient DNA have now been multiplied and analyzed from 40,000-year-old frozen mammoths, 18-million-year-old plant fossils, and 40-million-year-old insects entombed in amber. Progress in obtaining ancient human DNA has been slower, but Dr. Bryan Sykes of the Institute of Molecular Medicine, Oxford, collaborating with Chris Stringer, has succeeded in extracting DNA from the teeth of a 12,000-year-old Cro-Magnon jawbone from Gough's Cave, in England. The DNA includes a Y-chromosome segment, confirming that the individual concerned was a boy, and mitochondrial DNA that is closely related to that of living Europeans! And, in 1997, mitochondrial DNA was obtained from the actual Neander Valley skeleton of 1856, confirming that this, the type Neanderthal, really belonged to a separate lineage.[30] Despite the pessimistic words of Milford Wolpoff that the debate on modern human origins will go on forever because "that is the nature of science," a scientific resolution of the "Neanderthal problem," independent of any anatomical arguments, may be close at hand.

As we complete this book, further exciting evidence of the human revolution that propelled our original expansion around the globe is emerging in Europe, Asia, and Australia. In December 1994, one of the most spectacular painted caves yet seen was discovered near Avignon, in France. The Grotte Chauvet (named after the man who found it) has galleries covered in depictions of lions, bear, rhinos, horses, and deer. It also contains evidence of a cave bear shrine. Chauvet is clearly one of Europe's most significant painted caves, particularly as radiocarbon dating has revealed it to be between 30,000 and 34,000 years old.[31] It was therefore created by some of the first Cro-Magnons, at a time when Neanderthals were still clinging to their last European refuges, such as Zafarraya. The great sophistication of the Chauvet depictions means that most ideas of the stylistic evolution of such art must be abandoned. Similarly, dating techniques have revealed that around this time, Cro-Magnons were already colonizing the harsh landscapes around the Lena River in southern Siberia,[32] while on the other side of the world, their Australian counterparts were painting themselves with red ocher and producing necklaces of pierced shells very like those that adorned the bodies of their European kin. And in South America, a completely new and rich culture, contemporary with the Clovis people of North America, has been discovered.[33] All this work merely confirms our belief in the recent spread, out of Africa, of modern humans who brought their new marvels to an unsuspecting world.

Of course, the scientific impact of the Out of Africa theory has already been enormous. Ten years ago, despite the best efforts of researchers like Chris Stringer, Günter Bräuer, and Desmond Clark, it would not have been possible to organize a scientific congress to discuss our recent African Exodus—there were too few supporters and too many influential opponents. Today that's not the case; in fact its proponents are beginning to dominate the field. This change is reflected at every level, from grant proposals and fieldwork, teaching in universities, textbooks and popular books, to the content of novels and newspapers. Our African Exodus, once a heresy, is today's orthodoxy.

We therefore have many reasons to thank that man from Kibish who began our story. Two years after Chris last held the skull of that ancient forebear on the eve of its return to Ethiopia, Alex Haley—the black American author—told the story of the hunt for his own African "Roots" (as his book was called). In his narrative's emotional climax, he tells how, reunited with his lost Gambian kin, the village women pass him their babies in an ancient "laying on of hands" ceremony. Symbolically, Haley was being told: "Through this flesh, which is us, we are you and you are us."[34]

And so it is with the man from Kibish. He is humanity's African kin. He is us, and we are him.

Notes

1: The Kibish Enigma

Epigraph: G. B. Shaw, *Annajanska*.

1. J. H. Musgrave, 1973, "The phalanges of Neanderthal and Upper Palaeolithic hands," in M. Day (ed.), *Human Evolution*, Taylor and Francis: London, pp. 59–85.

2. C. Stringer, 1974, "Population relationships of later Pleistocene hominids: a multivariate study of available crania," *Journal of Archaeological Science*, 1:317–42.

3. R. Leakey, 1983, *One Life*, Michael Joseph: London.

4. R. Leakey: interview with R. McKie, 1993.

5. C. Stringer, 1978, "Some problems in middle and upper Pleistocene hominid relationships," in D. Chivers and K. Joysey (eds.), *Recent Advances in Primatology*, Academic Press: London, pp. 395–418.

6. M. H. Day and C. Stringer, 1982, "A reconsideration of the Omo Kibish remains and the *erectus-sapiens* transition," in *Proceedings of the 2nd International Congress of Human Palaeontology, Nice*, pp. 814–46.

7. R. Leakey, 1983, op. cit.

8. J. Shreeve, 1992, "The Dating Game," *Discover*, September.

9. J. Bronowski, 1973, *The Ascent of Man*, BBC: London.

2: East Side Story

Epigraph: Terry Pratchett, 1991, *Reaper Man*.

1. Darwin quoted from S. J. Gould, 1993, "Full of Hot Air," in *Eight Little Piggies*, Jonathan Cape: London, pp. 109–20.
2. C. Darwin, 1881, *The Descent of Man*, John Murray: London.
3. C. Linnaeus, 1758, *Systema Naturae*.
4. Russell, Lord Bertrand, 1872–1970.
5. P. Andrews, 1981, "Species diversity and diet in monkeys and apes during the Miocene," in C. B. Stringer (ed.), *Aspects of Human Evolution*, Taylor and Francis: London, pp. 25–61.
6. S. J. Gould, 1993, "The Declining Empire of Apes," in *Eight Little Piggies*, Jonathan Cape: London, pp. 284–95.
7. J. Diamond, 1991, *The Rise and Fall of the Third Chimpanzee*, Radius: London.
8. V. M. Sarich and A. C. Wilson, 1967, "Immunological time scale for hominid evolution," *Science* 158: 1200–1203.
9. P. Andrews and C. Stringer, 1993, "The Primates' Progress," in S. J. Gould (ed.), *The Book of Life*, Ebury-Hutchinson: London, pp. 219–51.
10. T. D. White, G. Suwa, and B. Asfaw, 1994, "*Australopithecus ramidus*, a new species of early hominid from Aramis, Ethiopia," *Nature*, 371: 306–12, and "*Ardipithecus*," 1995, *Nature*, 375: 88.
11. *Discover*, December 1994.
12. H. Gee, 1995, "New hominid remains found in Ethiopia," *Nature*, 373: 272.
13. A. Kortlandt, 1972, *New Perspectives on Ape and Human Evolution*, Stichting voor Psychobiologie: Amsterdam.
14. Y. Coppens, 1994, "East Side Story: the origin of Humankind," *Scientific American*, May: 62–69.
15. J. Reader, 1988, *Missing Links*, Penguin: Harmondsworth; R. Lewin, 1989, *Bones of Contention*, Penguin: Harmondsworth.
16. O. Lovejoy quoted in R. Leakey, 1994, *The Origin of Humankind*, Weidenfeld & Nicolson: London, p. 13.
17. R. Leakey, 1994, *The Origin of Humankind*, Weidenfeld & Nicolson: London, p. 13.
18. P. Wheeler: interview with R. McKie, 1994.
19. J. Reader, op. cit.; R. Lewin, op. cit.

20. R. Dart, 1953, "The predatory transition from ape to man," *International Anthropological and Linguistic Review*, Vol. 1, no. 4.

21. R. Ardrey, 1961, *African Genesis*, Collins: London.

22. C. K. Brain, 1981, *The Hunters or the Hunted?* University of Chicago Press: Chicago.

23. L. Berger and R. Clarke, (in press), "Eagle involvement in accumulation of the Taung child fauna," *Journal of Human Evolution*.

24. J. Reader, op. cit.; R. Lewin, op. cit.

25. The division between the Lower, Middle, and Upper Paleolithic was made in the nineteenth century. The Lower (now believed to have begun about 2.5 million years ago) was the earliest and simplest era of human technology. The Middle (sometimes also called the Mousterian, after a French Neanderthal site, and now taken as spanning the period from about 200,000 to 40,000 years ago) saw the introduction of a greater variety of tools. The Upper was the zenith of the Old Stone Age, characterized by the use of specialized tools, often made from blades; the first extensive working of bone, antler, and ivory; and the unequivocal presence of art.

26. C. Stringer, 1986, "The credibility of *Homo habilis*," in B. Wood, L. Martin, and P. Andrews (eds.), *Major Topics in Primate and Human Evolution*, Cambridge University Press: Cambridge, pp. 266–94; B. Wood, 1992, "Origin and evolution of the genus *Homo*," *Nature*, 355: 783–90.

27. J. Kingdon, 1993, *Self-Made Man and His Undoing*, Simon & Schuster: London.

28. L. Aiello: interview with R. McKie, 1995; L. Aiello and P. Wheeler, 1995, "The Expensive-Tissue Hypothesis," *Current Anthropology*, 36: 199–221.

29. J. Reader, op. cit.; R. Lewin, op. cit.

30. R. Leakey and R. Lewin, 1992, *Origins Reconsidered*, Little, Brown & Co: London, p. 34.

31. A. Walker and R. Leakey (eds.), 1994, *The Nariokotome Homo erectus Skeleton*, Springer-Verlag: Berlin.

32. L. Gabunia and A. Vekua, 1995, "A Plio-Pleistocene hominid from Dmanisi, East Georgia, Caucasus," *Nature*, 373: 509–12.

33. E. Carbonell and others, 1995, "Lower Pleistocene hominids and artifacts from Atapuerca—T.D. 6 (Spain)," *Science*, 269: 826–29.

34. M. Roberts, C. Stringer, and S. Parfitt, 1994, "A hominid tibia from Middle Pleistocene sediments at Boxgrove, UK," *Nature*, 369: 311–13.

35. J. Desmond Clark: quoted at the meeting in his honor, "The longest record: the human career in Africa," Berkeley, April 1986.

36. J. L. Arsuaga, I. Martinez, A. Gracia, J.-M. Carretero, and E. J.-L. Carbonell, 1993, "Three new human skulls from the Sima de los Huesos Middle Pleistocene site in Sierra de Atapuerca," *Nature*, 362: 534–37.

37. T. Li and D. Etler, 1992, "New Middle Pleistocene hominid crania from Yunxian in China," *Nature*, 357: 404–7.

38. G. J. Bartstra, S. Soegondho, and A. Wijk, 1988, "Ngandong Man: age and artifacts," *Journal of Human Evolution*, 17: 325–37.

39. J. Reader, op. cit.; R. Lewin, op. cit.

40. E. Trinkaus and P. Shipman, 1993, *The Neanderthals*, Knopf: New York; C. Stringer and C. Gamble, 1993, *In Search of the Neanderthals*, Thames & Hudson: London.

41. J. Reader, op. cit.; R. Lewin, op. cit.

42. J. Reader, op. cit.; R. Lewin, op. cit.

43. J. Radovčić, 1988, *Dragutin Gorjanović Kramberger and Krapina Early Man: The Foundation of Modern Palaeoanthropology*, Skolska knjiga and Hrvatski prirodoslovni muzej: Zagreb.

44. J. Pfeiffer, 1982, *The Creative Explosion*, Harper & Row: New York.

45. M. Boule and H. Vallois, 1946, *Les hommes fossiles*, Masson: Paris.

3: The Grisly Folk

Epigraph: C. L. Brace, 1964, "The fate of the 'classic' Neanderthals: a consideration of hominid catastrophism," *Current Anthropology*, 5: 3–43.

Epigraph: F. C. Howell quoted by C. Petit, *San Francisco Chronicle*, 13 February 1993.

1. J. H. Rosny-Aines, 1911, *La Guerre du Feu*, (reprinted as *Quest for Fire*, 1982, Penguin: Harmondsworth).

2. H. G. Wells, 1921, "The Grisly Folk," (reprinted in H. G. Wells, *Selected Short Stories*, 1958, Penguin: Harmondsworth).

3. E. Trinkaus and P. Shipman, 1993, *The Neanderthals*, Knopf: New York.

4. F. Weidenreich, 1940, "The Neanderthal Man and the ancestors of Homo sapiens," *American Anthropologist* 42: 375–383; 1942, "Facts and speculations concerning the origin of *Homo sapiens*," *American Anthropologist*, 49: 187–203; 1949, "Interpretations of the fossil material," *Studies in Physical Anthropology*, 1: 149–57.

5. C. S. Coon, 1962, *The Origin of Races*, Knopf: New York.

6. C. S. Coon, 1962, Book jacket quotations, op. cit.

7. T. Dobzhansky, 1963, "The Origin of the Races," *Scientific American*, February.

8. T. Dobzhansky, 1963, Letter to the Editor. *Scientific American*, April.

9. E. Trinkaus and P. Shipman, op. cit.

10. A. G. Thorne, 1980, "The centre and the edge: the significance of Australian hominids to African palaeoanthropology," in R. Leakey and B. Ogot (eds.), *Proceedings of the 8th Panafrican Congress of Prehistory and Quaternary Studies. Nairobi 1977*, The International Louis Leakey Memorial Institute for African Prehistory: Nairobi pp. 180–81

11. M. H. Wolpoff, Wu Xinzhi, and A. Thorne, 1984, "Modern *Homo sapiens* origins: a general theory of hominid evolution involving the fossil evidence from East Asia," in F. Smith and F. Spencer (eds.), *The Origins of Modern Humans: A World Survey of the Fossil Evidence*, Alan Liss: New York, pp. 411–83.

12. A. Thorne and M. Wolpoff, 1992, "The multiregional evolution of humans," *Scientific American*, April: 28–33.

13. A. Thorne and M. Wolpoff, 1991, "Conflict over modern human origins," *Search*, 22: 175–77.

14. G. P. Rightmire, 1989, *The evolution of Homo erectus*, Cambridge University Press: Cambridge.

15. W. Kimbel and L. Martin (eds.), 1993, *Species, Species Concepts, and Primate Evolution*, Plenum Press: New York.

16. J. Marks, 1994, *Discover*, November 1994.

17. C. Linnaeus, 1758, *Systema Naturae*.

18. J. F. Blumenbach, 1795, *De generis humani varietate nativa*, 3rd edition, Vandenhoek & Ruprecht: Gottingen.

19. C. S. Coon, 1962, op. cit.

20. C. Wills, 1992, "Has human evolution ended?" *Discover*, August: 22–24.

21. P. Frost, 1994, "Geographic distribution of human skin colour: a

selective compromise between natural selection and sexual selection?" *Human Evolution*, 9: 141–53.

22. M. Wolpoff et al., 1994, "The case for sinking *Homo erectus*: 100 years of *Pithecanthropus* is enough!" *Courier Forschungsinstitut Senckenberg*, 171: 341–61.

23. M. Boule and H. Vallois, 1946, *Les hommes fossiles*, Masson: Paris.

24. C. L. Brace, 1962, "Refocusing on the Neanderthal problem," *American Anthropologist*, 64: 729–41.

25. W. Golding, 1961, *The Inheritors*, Faber: London.

26. E. Trinkaus, 1983, *The Shanidar Neandertals*, Academic Press: New York.

27. R. S. Solecki, 1971, *Shanidar—The First Flower People*, Knopf: New York.

28. G. Constable, 1973, *The Neanderthals*. Time-Life Books: Amsterdam.

29. W. W. Howells, 1976, "Explaining modern man: evolutionists versus migrationists," *Journal of Human Evolution*, 5: 477–96.

30. P. Beaumont, H. de Villiers, and J. Vogel, 1978, "Modern man in sub-Saharan Africa present to 49,000 years BP: a review and evaluation with particular reference to Border Cave," *South African Journal of Science*, 74: 409–19.

31. R. Protsch, 1975, "The absolute dating of Upper Pleistocene sub-Sahara fossil hominids and their place in human evolution," *Journal of Human Evolution*, 4: 297–322.

32. T. McCown and A. Keith, 1939, *The Stone Age of Mount Carmel*, Vol. 2, Clarendon Press: Oxford.

33. A. Jelinek, 1982, "The Middle Palaeolithic in the Southern Levant with comments on the appearance of modern *Homo sapiens*," in A. Ronen (ed.), *The Transition from the Lower to the Middle Palaeolithic and the Origin of Modern Man*, British Archaeological Reports International Series, 151: Oxford pp. 57–104.

34. M. Aitken and H. Valladas, 1993, "Luminescence dating relevant to human origins," in M. Aitken, C. Stringer, and P. Mellars (eds.), op. cit.

35. R. Grün and C. Stringer, 1991, "Electron spin resonance dating and the evolution of modern humans," *Archaeometry*, 33: 153–99.

36. O. Bar-Yosef and B. Vandermeersch, 1993, "Modern humans in the Levant," *Scientific American*, April: 64–70.

37. E. Trinkaus, 1981, "Neanderthal limb proportions and cold adapta-

tion," in C. Stringer (ed.), *Aspects of Human Evolution*, Taylor and Francis: London, pp. 187–224.

38. C. Stringer, 1989, "The origin of early modern humans: a comparison of the European and non-European evidence," in P. Mellars and C. Stringer (eds.), *The Human Revolution*, Edinburgh University Press: Edinburgh, pp. 232–44.

39. G. Bräuer, 1984, "The Afro-European *sapiens* hypothesis, and hominid evolution in Asia during the Late Middle and Upper Pleistocene," in P. Andrews and J. Franzen (eds.), *The early evolution of Man, with special emphasis on Southeast Asia and Africa, Courier Forschungsinstitut Senckenberg*, 69: 145–66.

40. F. H. Smith: quoted in M. Brown, 1990, *The Search for Eve*, Harper-Collins: New York.

41. M. Wolpoff, 1989, "Multiregional evolution: the fossil alternative to Eden," in P. Mellars and C. Stringer (eds.), *The Human Revolution*, Edinburgh University Press: Edinburgh, pp. 62–108.

42. C. Stringer, 1984, "Ancestors: fate of the Neanderthal," *Natural History*, 93: 6–12.

43. C. L. Brace, 1994, review of "The origin of modern humans and the impact of chronometric dating," *Man*, 29: 473–75.

4: Time and Chance

Epigraph: L. P. Hartley, 1953, *The Go-Between*.

1. Y. Rak: interview with R. McKie at Amud, 1993.

2. Y. Rak, W. Kimbel, and E. Hovers, 1994, "A Neanderthal Infant from Amud Cave," *Journal of Human Evolution*, 26: 313–24. For an alternative view of these characters, see M. Creed-Miles, A. Rosas, and R. Kruszynski (in press), "Issues in the identification of Neanderthal derivative traits at early post-natal stages," *Journal of Human Evolution*.

3. H. Suzuki and F. Takai, 1970, *The Amud Man and His Cave Site*, University of Tokyo: Tokyo.

4. J. Shea: quoted in interview with D. Lieberman, Harvard, 1994.

5. B. Kurtén, 1979, "The shadow of the brow," *Current Anthropology*, 20: 229–30.

6. G. Krantz, 1973, "Cranial hair and brow ridges," *Mankind*, 9: 109–11.

7. J. Laitman, quoted in *Discover*, February 1992.

8. R. Franciscus and E. Trinkaus, 1988, "Nasal morphology and the emergence of *Homo erectus*," *American Journal of Physical Anthropology*, 75: 517–27.

9. J. Laitman, op. cit.

10. Y. Rak: interview with R. McKie at Amud, 1993.

11. B. Arensburg, L. Schepartz, A.-M. Tillier, B. Vandermeersch, H. Duday, and Y. Rak, 1990, "A reappraisal of the anatomical basis for speech in Middle Palaeolithic hominids," *American Journal of Physical Anthropology*, 83: 137–46; E. Culotta, 1993, "At each others' throats," *Science*, 260: 893; P. Lieberman, J. Laitman, J. Reidenberg, and P. Gannon, 1992, "The anatomy, physiology, acoustics and perception of speech," *Journal of Human Evolution*, 23: 447–67.

12. Quoted in R. Lewin, 1991, "Neanderthals puzzle the anthropologists," *New Scientist*, 20 April: 27.

13. A.J.E. Cave and W. L. Straus, 1957, "Pathology and posture of Neanderthal Man," *Quarterly Review of Biology*, 32: 348–63.

14. J. S. Jones, 1993, *The Language of the Genes*, HarperCollins: London.

15. O. Bar-Yosef and B. Vandermeersch, 1993, "Modern humans in the Levant," *Scientific American*, April: 64–70.

16. Y. Rak: interview with R. McKie at Amud, 1993.

17. J. Bronowski, 1973, *The Ascent of Man*, BBC: London.

18. J. Shea: quoted in interview with D. Lieberman, Harvard, 1994.

19. Y. Rak: interview with R. McKie at Amud, 1993.

20. A. Thorne and M. Wolpoff, 1992, "The multiregional evolution of humans," *Scientific American*, April: 28–33.

21. P. Mellars, 1995, quoted at the Royal Society/British Academy Meeting "Evolution of Social Behaviour Patterns in Primates and Man," London, April 1995.

22. D. Lieberman, 1993, "The rise and fall of seasonal mobility among hunter-gatherers: the case of the Southern Levant," *Current Anthropology*, 34: 599–631.

23. D. Lieberman: interview with R. McKie at Harvard, 1994.

24. L. Binford, quoted in *Discover* magazine, February 1992.

25. J-J Hublin: interview with R. McKie, at Zafarraya, 1994.

26. J-J. Hublin, C. Barroso Ruiz, P. Medina Lara, M. Fontugne, and J.-L Reyss, 1996, "The Mousterian site of Zafarraya (Andalucia, Spain): dating and implications on the Paleolithic peopling process of Western

Europe," *Compte Rendus de l'Académie des Sciences de Paris*, 321 (IIa): 931–37.

27. J. Kingdon, 1993, *Self-Made Man and His Undoing*, Simon & Schuster: London.

28. S. J. Gould, 1989, *Wonderful Life*, Hutchinson Radius: London.

29. E. Harth, 1990, *Dawn of a Millennium*, Penguin: London.

5: The Mother of All Humans?

Epigraph: J. Kingdon, 1993, *Self-Made Man and His Undoing*, Simon & Schuster: London.

 1. M. Ruvolo et al., 1993, "Mitochondrial COII Sequences and Modern Human Origins," *Molecular Biology and Evolution*, 10: 1115–35.

 2. A. Thorne and M. Wolpoff, *Scientific American*, op. cit.

 3. C. Wills, 1994, "Putting Human Genes on the Map," *Natural History*, June: 82–85.

 4. J. Watson, 1968, *The Double Helix*, Weidenfeld & Nicolson: London.

 5. W. Bodmer and R. McKie, 1994, *The Book of Man*, Little, Brown: London.

 6. R. Cann, M. Stoneking, and A. Wilson, 1987, "Mitochondrial DNA and Human Evolution," *Nature*, 325: 31–36.

 7. M. Brown, 1990, *The Search for Eve*, HarperCollins: New York.

 8. W. Bodmer and R. McKie, op. cit.

 9. M. Stoneking: interview with R. McKie, 1993, Birmingham, England.

10. V. Sarich and A. Wilson, 1967, "Immunological Timescale for Hominid Evolution," *Science*, 158: 1200–1203.

11. V. Sarich: quoted in R. Lewin, 1990, "Molecular Clocks Run out of Time," *New Scientist*, 10 February: 38–41.

12. L. Vigilant et al., 1989, "Mitochondrial DNA sequences in single hairs from a Southern African population," *Proceedings of the National Academy of Sciences, USA*, 86: 9350–4; L. Vigilant et al., 1991, "African populations and the evolution of human mitochondrial DNA," *Science*, 253: 1503–7.

13. S. Hedges et al., 1992, "Human origins and the analysis of mito-chondrial DNA sequences," *Science*, 255: 737–39.

14. A. Templeton, 1992, "Human origins and the analysis of mitochondrial DNA sequences," *Science*, 255: 737; A. Templeton, 1993, "The 'Eve'

Hypothesis: A genetic critique and re-analysis," *American Anthropologist*, 95: 51–72.

15. A. Templeton, publicity material prepared by Washington University, St. Louis.

16. M. Ruvolo: interview with R. McKie, 1994, Harvard.

17. C. Stringer and P. Andrews, 1988, "Genetic and fossil evidence for the origin of modern humans," *Science*, 239: 1263–68.

18. A. Merriwether et al., 1991, "The structure of human mitochondrial DNA variation," *Journal of Molecular Evolution*, 33: 543–55.

19. M. Stoneking: interview with R. McKie, 1993, Birmingham, England.

20. J. Kingdon, 1993, *Self-Made Man and His Undoing*, Simon & Schuster: London.

21. M. Ruvolo: interview with R. McKie, 1994, Harvard.

22. M. Hasegawa and S. Horai, 1991, "Time of the Deepest Root for Polymorphism in Human Mitochondrial DNA," *Journal of Molecular Evolution*, 32: 37–42.

23. M. Horai et al., 1995, "Recent African Origin of Modern Humans Revealed by Complete Sequences of Hominoid Mitochondrial DNAs," *Proceedings of the National Academy of Sciences, USA*, 92: 532–36.

24. K. Kidd and S. Tishkoff: interviews with C. Stringer and R. McKie, 1995.

25. M. Nei, 1995, "Genetic support for the Out of Africa theory of human evolution," *Proceedings of the National Academy of Science, USA*, 92: 6720–22.

26. W. Bodmer and R. McKie, 1994, *The Book of Man*, Little, Brown: London.

27. L. Cavalli-Sforza, P. Menozzi, and A. Piazza, 1994, *The History of Geography of Human Genes*, Princeton University Press: New Jersey.

28. L. Cavalli-Sforza, 1991, "Genes, Peoples and Languages," *Scientific American*, November 1991: 70–78.

29. S. Paabo, R. Dorit, and colleagues, "The Y chromosome and the origin of all of us (men)," 1995, *Science*, 268: 1141, 1183.

30. L. S. Whitfield, J. Sulston, and P. Goodfellow, 1995, "Sequence variation of the human Y chromosome," *Nature*, 378: 379–80.

31. M. Hammer, 1995, "A recent common ancestry for human Y chromosomes," *Nature*, 378: 376–78.

32. S. Pinker, 1994, *The Language Instinct*, Allen Lane: London.

33. J. S. Jones, 1993, *The Language of the Genes*, HarperCollins: London.
34. S. Pinker, 1994, *The Language Instinct*, op. cit.
35. L. Cavalli-Sforza et al., 1988, "Reconstruction of Human Evolution: Bringing together genetic, archaeological, and linguistic data," *Proceedings of the National Academy of Sciences, USA*, 85: 6002–6.
36. D. Penny, E. Watson, and M. Steel, 1993, "Trees from languages and genes are very similar," *Systematic Biology*, 42: 382–84.
37. A. Thorne and M. Wolpoff, 1992, "The Multiregional Evolution of Humans," *Scientific American*, April: 28–33.
38. H. Harpending et al., 1993, "Genetic structure of ancient human populations," *Current Anthropology*, 34: 483–96.
39. S. Rouhani, 1989, "Molecular genetics and the pattern of human evolution: plausible and implausible models," in P. Mellars and C. Stringer (eds.), *The Human Revolution*, Edinburgh University Press: Edinburgh, pp. 47–61.
40. L. Cavalli-Sforza, P. Menozzi, and A. Piazza, 1994, *The History and Geography of Human Genes*, op. cit.
41. Y. Rak: quoted in R. Lewin, 1991, "Neanderthals puzzle the anthropologists," *New Scientist*, 20 April: 27.
42. A. Gibbons, 1995, "Out of Africa—at Last?" *Science*, 267: 1272–73.
43. S. J. Gould, 1994, "So Near and Yet So Far," *New York Review of Books*: 24–28; "In the Mind of the Beholder," *Natural History*, February 1994: 14–23.

6: Footprints on the Sands of Time

Epigraph: H. W. Longfellow, A Psalm of Life.

1. Determination of the correct evolutionary position for the late archaic populations of China and Java is critical for both the Multiregional and Out of Africa models. For the multiregionalists, the path of evolution can be traced through the Beijing *Homo erectus* remains to Dali, and on to the 25,000-year-old early moderns of the Upper Cave at Zhoukoudian. But assessing the reality of these connections is not easy when some of the relevant fossils have been lost, and others are so zealously guarded by their local custodians. Nevertheless, on our reading, the archaic Chinese material is further away from modern Orientals in evolutionary terms than the equivalent evidence from Africa. And the

early modern people from sites like Omo, Skhul, and Qafzeh make better ancestors for the Upper Cave folk than do any preceding Chinese fossils. The multiregional evolutionary connection between Ngandong and Australia centers on an undated fossil braincase from an Australian region, the Willandra Lakes, which we will discuss later. Willandra Lakes Human 50 (WLH-50), as it is called, does have a skull form reminiscent of some Javanese fossils. Yet multivariate comparisons of shape carried out by Chris still link it more closely to African fossils like Jebel Irhoud, and even more so, to the Skhul and Qafzeh sample. So on our reading of the evidence, the archaic folk of the Far East were not the ancestors of the modern people who followed them.

2. M. Aitken, 1990, *Science-based Dating in Archaeology*, Longman: London.

3. J. Reader, 1988, *Missing Links*, Penguin: London.

4. J. Shreeve, 1992, "The Dating Game," *Discover*, September.

5. Quoted by J. Shreeve, 1992, "The Dating Game," op. cit.

6. Quoted in A. Gibbons, 1993, "Pleistocene population explosions," *Science*, 262: 27–28.

7. A. Gibbons, 1995, "The mystery of humanity's missing mutations," *Science*, 267: 35–36; A. Rogers and L. Jorde, 1995, "Genetic evidence of modern human origins," *Human Biology*, 67: 1–36.

8. C. B. Stringer, 1993, "Reconstructing recent human evolution," in M. Aitken, C. Stringer, and P. Mellars (eds.), *The Origin of Modern Humans and the Impact of Chronometric Dating*, Princeton University Press: New Jersey, pp. 179–95. J-J. Hublin, 1993, "Recent human evolution in Northwestern Africa," in op. cit. pp. 118–31. F. H. Smith, 1993, "Models and realities in modern human origins: the African fossil evidence," in op. cit. pp. 234–48. G. Bräuer, 1992, "Africa's place in the evolution of Homo sapiens," in G. Bräuer and F. Smith (eds.), *Continuity or Replacement: Controversies in Homo sapiens Evolution*, Balkema: Rotterdam pp. 83–98. G. P. Rightmire, 1989, "Middle Stone Age humans from Eastern and Southern Africa," in P. Mellars and C. Stringer (eds.), *The Human Revolution*, Edinburgh University Press, Edinburgh, pp. 109–22.

9. M. Rampino and S. Self, 1993, "Climate-volcanism feedback and the Toba eruption of ca. 74,000 years ago," *Quaternary Research*, 40: 269–80.

10. S. Ambrose, paper presented at the Palaeoanthropology Society Meetings, Anaheim, 1994.

11. H. Deacon, 1993, "Southern Africa and modern human origins," in M. Aitken, C. Stringer, and P. Mellars (eds.), *The Origin of Modern Humans and the Impact of Chronometric Dating*, op. cit., pp. 104–17; Beaumont, de Villiers, and Vogel, 1978, cited Chap. 3; C. Knight, C. Power, and I. Watts, 1995, "The human symbolic revolution: a Darwinian account," *Cambridge Archaeological Journal* 5: 75–114; A. Gibbons, 1995, "Old dates for modern behaviour," *Science*, 268: 495–96.

12. R. Klein, 1989, "Biological and behavioural perspectives on modern human origins in Southern Africa," in P. Mellars and C. Stringer (eds.), op. cit., pp. 529–46; S. Ambrose, 1993, paper presented at the Palaeoanthropology Society meetings, Anaheim, 1994.

13. M. Lahr and R. Foley, 1994, "Multiple dispersals and modern human origins," *Evolutionary Anthropology*, 3: 48–60.

14. J. Kingdon, 1993, *Self-Made Man and His Undoing*, Simon & Schuster: London.

15. C. B. Stringer, 1992, "Reconstructing recent human evolution," in M. Aitken, C. Stringer, and P. Mellars, op. cit., pp. 179–95.

16. C. Turner, 1992, "Microevolution of East Asian and European populations: a dental perspective," in T. Akazawa, K. Aoki, and T. Kimura (eds.), *The Evolution and Dispersal of Modern Humans in Asia*, Hokusen-Sha: Tokyo, pp. 415–38.

17. W. Broecker, 1994, "Massive iceberg discharges as triggers for global climatic change," *Nature*, 372: 421–24.

18. P. Ward, 1995, *The End of Evolution*, Weidenfeld & Nicolson: London.

19. C. Darwin, 1836: quoted in P. Ward, 1995, *The End of Evolution*, op. cit.

20. P. Martin: quoted in P. Ward, 1995, *The End of Evolution*, op. cit.

21. P. Ward, 1995, *The End of Evolution*, op. cit.

22. B. Kurtén: quoted in P. Ward 1995, op. cit.

23. P. Ward, 1995, op. cit.

24. J. Diamond, 1991, *The Rise and Fall of the Third Chimpanzee*, Radius: London.

25. P. Ward, 1995, *The End of Evolution*, op. cit.

26. C. Darwin: quoted in A. Desmond and J. Moore, 1991, *Darwin*, Michael Joseph: London.

27. L. Cavalli-Sforza, P. Menozzi, and A. Piazza, 1994, *The History and Geography of Human Genes*, Princeton University Press: New Jersey.

28. D. Wallace: quoted in A. Gibbons, 1993, "Geneticists trace the DNA

trail of the first Americans," *Science*, 259: 312–13. For an alternative view see A. Gibbons, 1996, "The peopling of the Americas," *Science*, 274: 31–33.

29. D. Metzer, 1995, "Stones of contention," *New Scientist*, 24 June: 31–35.
30. J. Flood, 1989, *Archaeology of the Dreamtime*, Collins: Australia.
31. Ibid.
32. A. Thorne, 1977, "Separation or reconciliation? Biological clues to the development of Australian society," in J. Allen, J. Golson, and R. Jones (eds.), *Sunda and Sahul*, Academic Press: London, pp. 187–204.
33. P. Brown, 1987, "Pleistocene homogeneity and Holocene size reduction: the Australian human skeletal evidence," *Archaeology in Oceania*, 22: 47–71.
34. P. Brown, 1994, "Cranial vault thickness in Asian *Homo erectus* and *Homo sapiens*," *Courier Forschungsinstitut Senckenberg*, 171: 33–46.
35. C. Pardoe, 1993, "Competing paradigms and ancient human remains: the state of the discipline," *Archaeology in Oceania*, 26: 79–85.
36. R. Sim and A. Thorne, 1990, "Pleistocene human remains from King Island, Southeastern Australia," *Australian Archaeology*, 31: 44–51; P. Brown, 1994, "A flawed vision: sex and robusticity on King Island," *Australian Archaeology*, 38: 1–7; A. Thorne and R. Sim, 1994, "The gracile male skeleton from Late Pleistocene King Island, Australia," *Australian Archaeology*, 38: 8–10.
37. J. Diamond, 1991, *The Rise and Fall of the Third Chimpanzee*, op. cit.
38. R. G. Roberts, R. Jones, and M. Smith, 1994, "Beyond the Radiocarbon barrier in Australian Prehistory," *Antiquity*, 68: 611–16.
39. J. Kingdon, 1993, *Self-Made Man and His Undoing*, Simon & Schuster: London.

7: Africans Under the Skin

Epigraph: E. Harth, 1990, *Dawn of a Millennium*, Penguin: London.
Epigraph: B. Okri, 1995, *Astonishing the Gods*, Phoenix House: London.
1. P. Mitchell, "Africa and the West in historical perspective," quoted in R. Coughlan, 1963, *Tropical Africa*, Time-Life International: Netherlands, p. 109. We are grateful to Mr. R. Snelling for this and the succeeding quotation.

2. G. Dieterlen, 1975, Introduction to M. Griaule, *Conversations with Ogotemmeli*, Oxford University Press: New York, p. xiv.

3. A. Gibbons, 1995, "Old dates for modern behaviour," *Science*, 268: 495–96; A. Brooks et al., 1995, "Dating and context of three Middle Stone Age sites with bone points in the Upper Semliki Valley, Zaire," *Science*, 268: 548–53; J. Yellen et al., 1995, "A Middle Stone Age worked bone industry from Katanda, Upper Semliki Valley, Zaire," *Science*, 268: 553–56.

4. A. Rogers and L. Jorde, 1995, "Genetic evidence on modern human origins," *Human Biology*, 67: 1–36.

5. R. Lewontin, 1982, *Human Diversity*, Scientific American Library: New York.

6. A. Roger and L. Jorde, 1995, op. cit.; J. Mountain and L. Cavalli-Sforza, 1994, "Inference of human evolution through cladistic analysis of nuclear DNA restriction polymorphisms," *Proceedings of the National Academy of Sciences USA*, 91: 6515–19.

7. J. Relethford and H. Harpending, 1994, "Craniometric variation, genetic theory, and modern human origins," *American Journal of Physical Anthropology*, 95: 249–70.

8. J. P. Rushton, "Evolutionary biology and heritable traits," Paper presented at the Annual Meeting of the American Association for the Advancement of Science, San Francisco, 1989.

9. J. P. Rushton, 1997, *Race, Evolution, and Behavior*, Transaction Publishers: New Brunswick & London.

10. M. Kohn, 1995, *The Race Gallery*, Cape: London.

11. R. J. Herrnstein and C. Murray, 1994, *The Bell Curve: Intelligence and Class Structure in American Life*, The Free Press: New York.

12. *The New Republic*, 31 October 1994.

13. A. Fausto-Sterling, 1993, "Sex, race, brains and calipers," *Discover*, October: 32–37.

14. R. Martin and K. Saller, 1956, *Lehrbuch der Anthropologie*, Gustav Fischer: Stuttgart.

15. K. L. Beals, C. L. Smith, and S. M. Dodd, 1984, "Brain size, cranial morphology, climate and time machines," *Current Anthropology*, 25: 301–30.

16. J. P. Rushton, 1995, "Race and crime: an international dilemma," *Society*, 32: 37–41.

17. C. Darwin: quoted in A. Desmond and J. Moore, 1991, *Darwin*, Michael Joseph: London.

18. M. Henneberg, A. Budnik, M. Pezacka, and A. E. Puch, 1985, "Head size, body size, and intelligence: intraspecific correlations in *Homo sapiens sapiens*," *Homo*, 36: 207–18.

19. Davidson Nicol obituary, *The Times* (London), 19 October 1994.

20. "Africans move to the top of Britain's education ladder," Sunday Times (London), 23 January 1994.

21. T. Beardsley, 1995, "For whom the Bell Curve really tolls," *Scientific American*, January: 8–10.

22. J. C. Gutin, 1994, "End of the rainbow," *Discover*, November: 71–75.

23. Quoted in "Race and Color," *Discover*, November: 82–89.

24. J. Diamond, 1991, *The Rise and Fall of the Third Chimpanzee*, Radius: London.

25. Quoted in "Race and Color," *Discover*, November: 82–89.

26. J. Swerdlow, 1995, "Quiet miracles of the brain," *National Geographic*, June: 2–41.

8: The Sorcerer

Epigraph: R. Dawkins, 1976, *The Selfish Gene*, Oxford University Press: Oxford.

1. J. Pfeiffer, 1982, *The Creative Explosion*, Harper & Row: New York.

2. L. Binford, 1989, "Isolating the transition to cultural adaptations: an organizational approach," in E. Trinkaus (ed.), *The Emergence of Modern Humans: Bioculture Adaptations in the Later Pleistocene*, Cambridge University Press: Cambridge, pp. 18–41.

3. W. Calvin, 1994, "The emergence of intelligence," *Scientific American*, October.

4. L. Cosmides, quoted by W. Allman, 1994, *The Stone Age Present*, Simon & Schuster: New York.

5. C. Lumsden and E. O. Wilson, 1983, *Promethean Fire: Reflections on the Origin of Mind*, Harvard University Press: Cambridge, Mass.

6. L. Cosmides and J. Tooby, "The lords of many domains," *The Times* (London) *Higher Educational Supplement*, 25 June 1993.

7. R. Wright, 1994, *The Moral Animal: Why We Are the Way We Are*, Pantheon: New York.

8. L. Cosmides: interview with R. McKie, 1995.

9. C. Darwin, quoted by S. Pinker, "The Language Instinct," *The Times* (London) *Higher Educational Supplement*, 25 June 1993.

10. Ibid.

11. D. B. Fry.

12. Robin Dunbar, *The Times* (London), 5 February 1994.

13. S. Pinker, 1994, *The Language Instinct*, Allen Lane: London.

14. K. Kidd: interview with R. McKie at Yale, 1992.

15. J. Goodall, 1990, *Through a Window*, Penguin: London.

16. P. Lieberman, J. Laitman, J. Reidenberg, and P. Gannon, 1992, "The anatomy, physiology, acoustics and perception of speech," *Journal of Human Evolution*, 23: 447–67.

17. B. Arensburg, L. Schepartz, A.-M. Tillier, B. Vandermeersch, H. Duday, and Y. Rak, 1990, "A reappraisal of the anatomical basis for speech in Middle Palaeolithic hominids," *American Journal of Physical Anthropology*, 83: 137–46; E. Culotta, 1993, "At each others' throats," *Science*, 260: 893.

18. R. Dunbar: interview with R. McKie, 1994; R. Dunbar, "The Chattering Classes: what separates us from the animals," *The Times* (London), 5 February 1994; Gail Vines, "Essential Chatlines—people are all talk," Guardian (London), 7 August 1992.

19. L. Binford: quoted in J. Fischman, 1992, "Hard Evidence," *Discover*, February.

20. S. Mithen, paper presented at the Royal Society/British Academy Meeting "Evolution of Social Behaviour Patterns in Primates and Man," London, April 1995.

21. D. Vialou: quoted in R. Lewin, 1993, *The Origin of Modern Humans*, Scientific American Library: New York.

22. J. Marshack, 1991, *The Roots of Civilisation*, Moyer Bell: New York.

23. C. Gamble: interview with R. McKie, 1995.

24. R. White, 1993, "The dawn of adornment," *Natural History*, May, 1193: 60–67.

25. R. White, 1993, "Technological and social dimensions of 'Aurignacian-age' body ornaments across Europe," in H. Knecht, A. Pike-Tay, and R. White (eds.), *Before Lascaux*, CRC Press: Boca Raton, pp. 277–99.

26. I. DeVore: quoted by W. Allman, 1994, op. cit.

27. W. Allman, 1994 op. cit.

28. D. Buss, 1994, "The strategies of human mating," *American Scientist*, 82: 238–49.

29. M. Ridley, 1993, *The Red Queen: Sex and the Evolution of Human Nature*, Penguin: London.

30. C. Knight, 1991, *Blood Relations: Menstruation and the Origins of Culture*, Yale University Press: New Haven; C. Knight, C. Power, and I. Watts, 1995, "The human symbolic revolution: a Darwinian account," *Cambridge Archaeological Journal*, 5: 75–114.

31. O. Soffer, 1994, "Ancestral lifeways in Eurasia—the Middle and Upper Paleolithic records," in M. and D. Nitecki (eds.), *Origins of Anatomically Modern Humans*, Plenum Press: New York, pp. 101–19.

32. L. Binford, 1989, "Isolating the transition to cultural adaptations: an organizational approach," in E. Trinkaus (ed.), *The Emergence of Modern Humans: Biocultural Adaptations in the Later Pleistocene*, Cambridge University Press: Cambridge, pp. 18–41. (See also L. Binford, quoted in *Discover* magazine, February 1992.)

9: Prometheus Unbound

Epigraph: A. Carter, quoted in *Focus* magazine.
Epigraph: B. Russell, 1952, *Impact of Science on Society*, Chapter 7.

1. D. Pilbeam: quoted by R. Leakey and R. Lewin in *Origins Reconsidered*, 1992, Little, Brown: London.

2. A. Ballantyne, 1994, "The wisdom or folly of pulling teeth," *The Times* (London).

3. S. J. Gould, 1980, "Our Greatest Evolutionary Step," *Panda's Thumb*, Penguin: London.

4. L. Aiello: interview with R. McKie, 1995.

5. R. Lewin, 1995, "Rise and fall of big people," *New Scientist*, 26 April: 30–33.

6. P. Brown: interview with R. McKie, 1995.

7. R. Lewin, op. cit.

8. J. Kingdon, 1993, *Self-Made Man and His Undoing*, Simon & Schuster: London.

9. T. Molleson, 1994, "The eloquent bones of Abu Hureyra," *Scientific American*, August: 60–65; T. Molleson: interview with R. McKie, 1994.

10. J. Diamond, 1991, *Rise and Fall of the Third Chimpanzee*, Radius: London.

11. R. Nesse and G. Williams, 1994, *Why We Get Sick*, Times Books: New York.

12. R. Di Silvestro: quoted in P. Ward, 1995, *The End of Evolution*, Weidenfeld & Nicolson: London.

13. Friends of the Earth, 1995.

14. R. Buddemeier: quoted in P. Ward, 1995, op. cit.

15. E. Harth, 1990, *Dawn of a Millennium*, Penguin: London.

16. J. S. Jones, 1994, "A Brave, New, Healthy World?," *Natural History*, June: 72-85.

17. C. Wills, 1992, "Has Human Evolution Ended?" *Discover*, August: 22–24.

18. Tu-Wa-Moja is Swahili for "We are one." The problems of the Museum of Natural History in Washington, D.C., were recounted in J. Achenbach, 1991, "Little White Lies," *Washington Post*, 13 October 1991.

19. B. Ortiz de Montellano, 1993, "Melanin, afrocentricity, and pseudo-science," *Yearbook of Physical Anthropology*, 36: 33–58.

20. M. Leakey, 1995, "The farthest horizon," *National Geographic*, September.

21. J. Shreeve, 1996, "New skeleton gives path from trees to ground an odd turn," *Science*, 272: 654.

22. E. Carbonell et al., 1995, "Lower Pleistocene hominids and artifacts from Atapuerca-TD6 (Spain)," *Science*, 269: 826–29; J. Bermudez de Castro et al., 1997, "A hominid from the Lower Pleistocene of Atapuerca, Spain: possible ancestor to Neanderthals and modern humans," *Science* 276: 1392–95.

23. V. Pesce-Delfino and E. Vacca, 1993, "An archaic human skeleton discovered at Altamura (Bari, Italy)," *Rivista di Antropologia*, 71: 249–57.

24. T. Akazawa et al., 1995, "Neanderthal infant burial," *Nature*, 377: 585–86.

25. E. Vacca and D. Coppola, 1993, "The Upper Paleolithic burials at the cave of Santa Maria di Agnano near Ostuni (Brindisi, Italy): preliminary report," *Rivista di Antropologia*, 71: 275–84.

26. C. Zollikofer et al., 1995, "Neanderthal computer skulls," *Nature*, 375: 283–85.

27. J-J. Hublin, F. Spoor, et al., 1996, "A late Neanderthal associated with Upper Palaeolithic artefacts," *Nature*, 381: 224–26.

28. M. Schoeninger, 1995, "Stable isotope studies in human evolution," *Evolutionary Anthropology*, 4: 83–98.

29. S. Paabo, 1993, "Ancient DNA," *Scientific American*, November.
30. M. Krings et al., 1997, "Neanderthal DNA sequences and the origin of modern humans," *Cell* 90: 19–30; R. Ward & C. Stringer 1997, "A molecular handle on the Neanderthals," *Nature* 388: 225–26.
31. M. Lemonick, 1995, "Stone-Age Bombshell," *Time*, 19 June 1995.
32. V. Morell, 1995, "Siberia: surprising home for early modern humans," *Science*, 268: 1279.
33. J. Hecht, 1996, "You take the coast road . . ." *New Scientist*, 27 April: 21.
34. A. Haley, 1977, *Roots*, Pan: London.

Index

Page numbers in italics refer to illustrations.